chemistry, electron microscopy, macroscopy, experimental morphology and embryology and comparative anatomy) are published in Advances in Anatomy, Embryology and Cell Biology. Papers dealing with anthropology and clinical morphology that aim to encourage co-operation between anatomy and related disciplines will also be accepted.

Papers are normally commissioned. Original papers and communications may be submitted and will be considered for publication provided they meet the requirements of a review article and thus fit into the scope of "Advances". English language is preferred, but in exceptional cases French or German papers will be accepted.

It is a fundamental condition that submitted manuscripts have not been and will not simultaneously be submitted or published elsewhere. With the acceptance of a manuscript for publication, the publisher acquires full and exclusive copyright for all languages and countries.

Twenty-five copies of each paper are supplied free of charge.

Manuscripts should be addressed to

Prof. Dr. F. **BECK,** Department of Anatomy, University of Leicester, 6 University Road, GB-Leicester LE1 7RH

Prof. W. **HILD,** Department of Anatomy, Medical Branch, The University of Texas, Galveston, Texas 77550/USA

Prof. Dr. R. **ORTMANN,** Anatomisches Institut der Universität, Lindenburg, D-5000 Köln-Lindenthal

Prof. J.E. **PAULY,** Department of Anatomy, University of Arkansas for Medical Sciences, Little Rock, Arkansas 72205/USA

Prof. Dr. T.H. **SCHIEBLER,** Anatomisches Institut der Universität, Koellikerstraße 6, D-8700 Würzburg

Advances in Anatomy
Embryology and Cell Biology

Vol. 86

Editors
F. Beck, Leicester W. Hild, Galveston
R. Ortmann, Köln J.E. Pauly, Little Rock
T.H. Schiebler, Würzburg

Jeffery A. Winer

The Medial Geniculate Body of the Cat

With 45 Figures

Springer-Verlag
Berlin Heidelberg NewYork Tokyo
1985

Jeffery A. Winer
Assistant Professor of Anatomy
Department of Physiology-Anatomy
University of California
Berkeley, California 94720 U.S.A.

Library of Congress Cataloging in Publication Data
Winer, Jeffery A., 1945–. The medial geniculate body of the cat.
(Advances in anatomy, embryology, and cell biology; vol. 86)
Bibliography: p. Includes index.
1. Medial geniculate body. 2. Cats – Anatomy. 3. Mammals – Anatomy.
I. Title. II. Series: Advances in anatomy, embryology, and cell biology; v. 86.
QL801.E67 vol. 86a   574.4s   [599.74′428]   84-5361   [QL938.M43]
ISBN-13: 978-3-540-13254-7   e-ISBN-13: 978-3-642-69634-3
DOI: 10.1007/978-3-642-69634-3

The use of general descriptive names, trade names, trade marks, etc. in this
publication, even if the former are not especially identified, is not to be
taken as a sign that such names, as understood by the Trade Marks and
Merchandise Marks Act, may accordingly be used freely by anyone. Pro-
duct Liability: The publisher can give no guarantee for information about
drug dosage and application thereof contained in this book. In every individual case the respective user must check its accuracy by consulting other
pharmaceutical literaure.

2121/3140-543210

Dedicated to my Mother and Father

Preface

In the operation of reasoning, the mind does nothing but run over its objects, as they are supposed to stand in reality, without adding any thing to them or diminishing any thing from them. If I examine the Ptolomaic and Copernican systems, I endeavour only, by my inquiries, to know the real situation of the planets; that is, in other words, I endeavour to give them, in my conception, the same relation that they bear towards each other in the heavens. To this operation of the mind, therefore, there seems to be always a real, though often unknown standard, in the nature of things; nor is truth or falsehood variable by the various apprehensions of mankind.

D. Hume, The sceptic.
In: Essays. Moral Political and Literary.
Oxford University Press, Oxford, 1963, p. 166.

Contents

Abbreviations

A,AAF	anterior auditory field
aeg	anterior ectosylvian gyrus
aes	anterior ectosylvian sulcus
Aq	cerebral aqueduct
AI	primary auditory cortical area
AII	second auditory cortical area
AIII	third auditory cortical area
BIC	brachium of the inferior colliculus
BICv	ventral nucleus of the brachium of the inferior colliculus
BSC	brachium of the superior colliculus
BV	blood vessel
CG	central gray substance
CGL	lateral geniculate body
CM	centre-médian nucleus
CP	cerebral peduncle
D	dorsal nucleus of the dorsal division of the medial geniculate body
Da	anterior dorsal nucleus of the dorsal division of the medial geniculate body
DD	deep dorsal nucleus of the dorsal division of the medial geniculate body
DDa	anterior deep dorsal nucleus of the dorsal division of the medial geniculate body
DS	superficial dorsal nucleus of the dorsal division of the medial geniculate body
DSa	anterior superficial dorsal nucleus of the dorsal division of the medial geniculate body
EE	excitatory-excitatory binaural cortical band
EI	excitatory-inhibitory binaural cortical band
Ep	posterior ectosylvian cortex
G	glial cell(s)
Ha	habenula
HaI	habenulo-interpeduncular tract
I	inferior nucleus of pulvinar
IBIC	interstitial nucleus of the brachium of the inferior colliculus
IC	inferior colliculus
Ins	insular cortical field
L	posterior limitans nucleus of the dorsal division
lat	lateral gyrus
LI	lateral incisure of the medial geniculate body *or* lateral inferior thalamic nucleus
LMN	lateral mesencephalic nucleus

LP	lateral posterior nucleus
LPc	caudal division of the lateral posterior nucleus
M	medial (magnocellular division) of the medial geniculate body
MB	mammillary body
meg	middle ectosylvian gyrus
MRF	mesencephalic recticular formation
mss	middle suprasylvian sulcus
MZ	marginal zone of the medial geniculate body
NOT	nucleus of the optic tract
NIII,rNIII	oculomotor nerve *or* root
OR	optic radiation
OT	optic tract
OV	ovoid nucleus of the ventral division of the medial geniculate body
P	posterior field of the ectosylvian gyrus
PC	posterior commissure
peg	posterior ectosylvian gyrus
pes	posterior ectosylvian sulcus
psb	perisomatic basket
pss	pseudosylvian sulcus
Pt	pretectal nuclei
PUL,Pul	pulvinar nuclei
rh	rhinal sulcus
RN	red nucleus
SC	superior colliculus
sf	suprasylvian fringe cortex
SG	suprageniculate nucleus of the dorsal division of the medial geniculate body
SNL	lateral division of substantia nigra
SP	suprapeduncular nucleus
SPF,SPN	subparafascicular nucleus
Te	temporal cortical field
V	ventral nucleus of the ventral division of the medial geniculate body
VB	ventrobasal complex
VGL	ventral nucleus of the lateral geniculate body
VL	ventrolateral nucleus of the dorsal division of the medial geniculate body
VP	ventroposterior field of the posterior ectosylvian gyrus
3rd vent	third ventricle
Orientation of section	A, anterior; D, dorsal; L, lateral; M, medial; P, posterior; V, ventral

1 Introduction

The medial geniculate body is the obligatory synaptic relay in the thalamus for ascending auditory information from many brain stem nuclei. The main source is the inferior colliculus (Kudo and Niimi 1980), but other sources include the lateral tegmental system of the midbrain (Morest 1965b) and the superior olivary complex (Papez 1929a; Henkel 1983). The medial geniculate body is also a primary target for descending fibers from the cerebral cortex (Diamond et al. 1969) and has a major input to each subdivision of the auditory cortex (Winer et al. 1977; Niimi and Matsuoka 1979) as well as lesser projections to other parts of the cerebral cortex (Spreafico et al. 1981). Interposed between ascending midbrain and descending cortical auditory (and non-auditory) influences, the position, size, and complexity of the medial geniculate body suggest a pivotal and diversified role in hearing. Besides its geographical centrality as a synaptic station for confluent ascending and descending influences, the medial geniculate body also functions as an integrative center with its own complement of interneurons and intrinsic synaptic architecture (Morest 1965a, 1975a; Winer 1979; Winer and Morest 1978, 1983a, b, 1984).

While the medial geniculate body can be regarded as an entity whose function is principally auditory, to consider it as exclusively so ignores the diversity and complexity of parallel channels for sensory information processing (Moore and Goldberg 1963; Erulkar 1975; Casseday et al. 1976; Oliver and Hall 1978a, b). Such channels provide multiple central neural representations of individual sensory epithelia in the cerebral cortex and have now been described for the visual (Palmer et al. 1978; Tusa et al. 1978), somatic sensory (Merzenich et al. 1978), and auditory (Woolsey 1961; Reale and Imig 1980) systems. Analysis of the morphological bases for these multiple channels and the intrinsic organization of the nuclei in the medial geniculate body is a goal of this paper. This precedes and complements a functional architecture and is a prelude to a more precise delineation of the parallel or convergent auditory and non-auditory pathways in the medial geniculate body.

The structure of nerve cells in the auditory system has been used previously as a first approximation of function (Kiang et al. 1973; Morest et al. 1973; Cant and Morest 1979; Bourk et al. 1981). In the same spirit morphological data of the present study could serve as physiological predictors. Neural structure is also a critical adjunct for the study of homology (Morest 1965c; Winer and Morest 1979; Morest and Winer to be published). Since a thalamic relay nucleus for audition is found in mammals (Ariëns Kappers et al. 1936), reptiles (Pritz 1980), and birds (Karten 1967), it is of interest to find if a common structural plan (Kuhlenbeck 1966, 1973) applies to the mammalian medial geniculate body. The data reported here are an amalgam of old and new results whose provenance is identified in each case and in which morphological and connectional approaches are integrated.

2 Materials and Methods

The brain stems of about 250 cats ranging in age from newborn to adult were available for study. A more detailed account of the various methods is available (Winer et al. 1977; Winer 1984a, b; Winer and Morest 1983a, b, 1984). Some 200 animals were stained using the Golgi method. Of these, the majority were impregnated using the rapid Golgi technique (Golgi 1878, 1879, 1891; Morest and Morest 1966). Other animals from this series were perfused intracardially with formalin or mixed aldehydes and stained using the Golgi-Kopsch method (Kopsch 1896). About 40 brain stems were impregnated according to Cox's Golgi protocol (Cox 1891; Golgi 1891; Ramon-Moliner 1970). For these staining technics thick serial sections (100–250 μm) were made in the cardinal planes or oblique to these from material embedded in low-viscosity nitrocellulose.

A smaller number of adult animals (about 30) were perfused with formalin or mixed aldehydes and the brain embedded in celloidin, low-viscosity nitrocellulose, or paraffin, sectioned serially, and stained for Nissl substance. The brain stem was usually sectioned in the stereotaxic transverse plane or parallel to the floor of the fourth ventricle. Some sections were stained for fibers with either Bodian's (Bodian 1937), Weil's, Woelcke's, or Weigert's method and the fiber plexus from these referenced to the cytoarchitectonic boundaries as defined from Nissl and Golgi material. The medial geniculate body from eight adult animals was prepared for electron microscopy and embedded in Epon or Araldite. For light microscopy, 0.5- to 2-μm-thick sections were stained with toluidine blue.

Other animals received injections of various tracer substances, usually horseradish peroxidase (LaVail et al. 1973; Winer 1977) or tritiated leucine (Cowan et al. 1972) for connectional studies of the medial geniculate body. Frozen sections from these experiments, stained for cells or fibers, were available for study.

About 250 serially sectioned brain stems from other species and containing the medial geniculate body or its presumptive homologue were available. This included Golgi-impregnated material from primates (*Macaca mulatta, Papio papio, Homo sapiens*), insectivores (*Antrozous pallidus* and *Pteronotus p. parnellii*), marsupials (*Didelphis marsupialis virginiana*), rodents (*Sciurus carolinensis* and *Rattus norvegicus*), and birds (*Gallus gallus*). Unless otherwise noted, however, the observations in the present account are limited to the cat.

Material from connectional and morphological experiments was studied with semi-, planapo-, or planachromatic, oil-immersion objectives with high numerical aperture (1.05–1.4) and long working distance (300 μm or more). The lens used for each drawing and its numerical aperture (N.A.) is indicated. Most drawings were made at a final magnification of × 1250, and others at final magnifications ranging from × 40 to × 2250 for, respectively, low-power survey work or to render complex detail.

The following definitions are used to describe the neuronal types in this study.

The terminology conforms to patterns with a long tradition in neuroanatomy (Kölliker 1896; Ramón y Cajal 1911; Lorente de Nó 1934). Dendritic branching is described as *bushy* if the dendrites form parallel, richly branched tufts with a more or less distinct polarity, or as *stellate* if their branches radiate to fill a sphere and divide more obliquely. Since the Golgi methods are incapable of demonstrating synaptic relations, the use of the terms "ending" and "target" implies only that certain forms of apposition are repeatedly observed. The three-dimensional shape of the dendritic field ranges from spherical to polarized to cylindrical or conelike.

3 Observations

3.1 Topography of the Medial Geniculate Body

The medial geniculate body forms the posterolateral extremity of the diencephalon. In the transverse plane its caudal, free part protrudes toward the mesencephalon, and the superior colliculus and midbrain reticular formation form, respectively, its dorsal and medial borders. Laterally, the hippocampal formation surrounds it. Rostrally, the pulvinar and associated nuclei form the dorsomedial border, the lateral geniculate body is the dorsolateral margin, and the subparafascicular and suprapeduncular nuclei abut the ventral border. Sections through the rostral quarter of the medial geniculate body (Fig. 2D) show the fibers of the optic tract on its lateral border, lateral lemniscal fibers ventromedially, and, still more medially, medial lemniscal axons en route to the ventrobasal complex. At this and more rostral levels, the medial geniculate body is penetrated by large contingents of thalamofugal and corticofugal fibers, axons from the brachium of the inferior colliculus, and intrinsically arborizing fibers, which, together, form an exceedingly dense neuropil in which the anterior tier of nuclei are embedded. The anterior nuclei give way to the transitional zone intercalated between the rostral pole of the medial geniculate body and the caudal part of the ventrobasal complex (Spreafico et al. 1981; Winer and Morest 1983a).

3.2 Cytoarchitectonic Subdivisions of the Medial Geniculate Body

The features of dendritic architecture which distinguish the ventral, dorsal, and medial divisions — the three main parts of the medial geniculate body (Morest 1964, 1965a) — are apparent in Golgi preparations (Fig. 1A). Neurons are differentiated into type I (projection) or type II (interneuronal) cells on the bases of size, dendritic and axonal configuration, and patterns of connections (Golgi 1906).

The major divisions are observed also in Nissl-stained material (Figs. 3, 4). The *ventral division* consists of medium-sized and small cells (Fig. 4A, C; Fig. 44A, B), which are densely packed and form long rows. The dendrites of the larger neurons are polarized along the dorsal-to-ventral axis in Golgi

Fig. 1A, B. Major subdivisions of the medial geniculate body of the cat. A Transverse recon- ▷ struction of two serial, 160-μm-thick sections at the caudal third of the medial geniculate body to show the characteristic neuronal architecture. Golgi-Cox method, adult cat; planachromat, N.A. 0.35, ×200. B Axonometric view of the major nuclei of the medial geniculate body from an anterolateral perspective after dissection of the overlying optic tract and lateral thalamic nuclei. (From Winer and Morest 1983a)

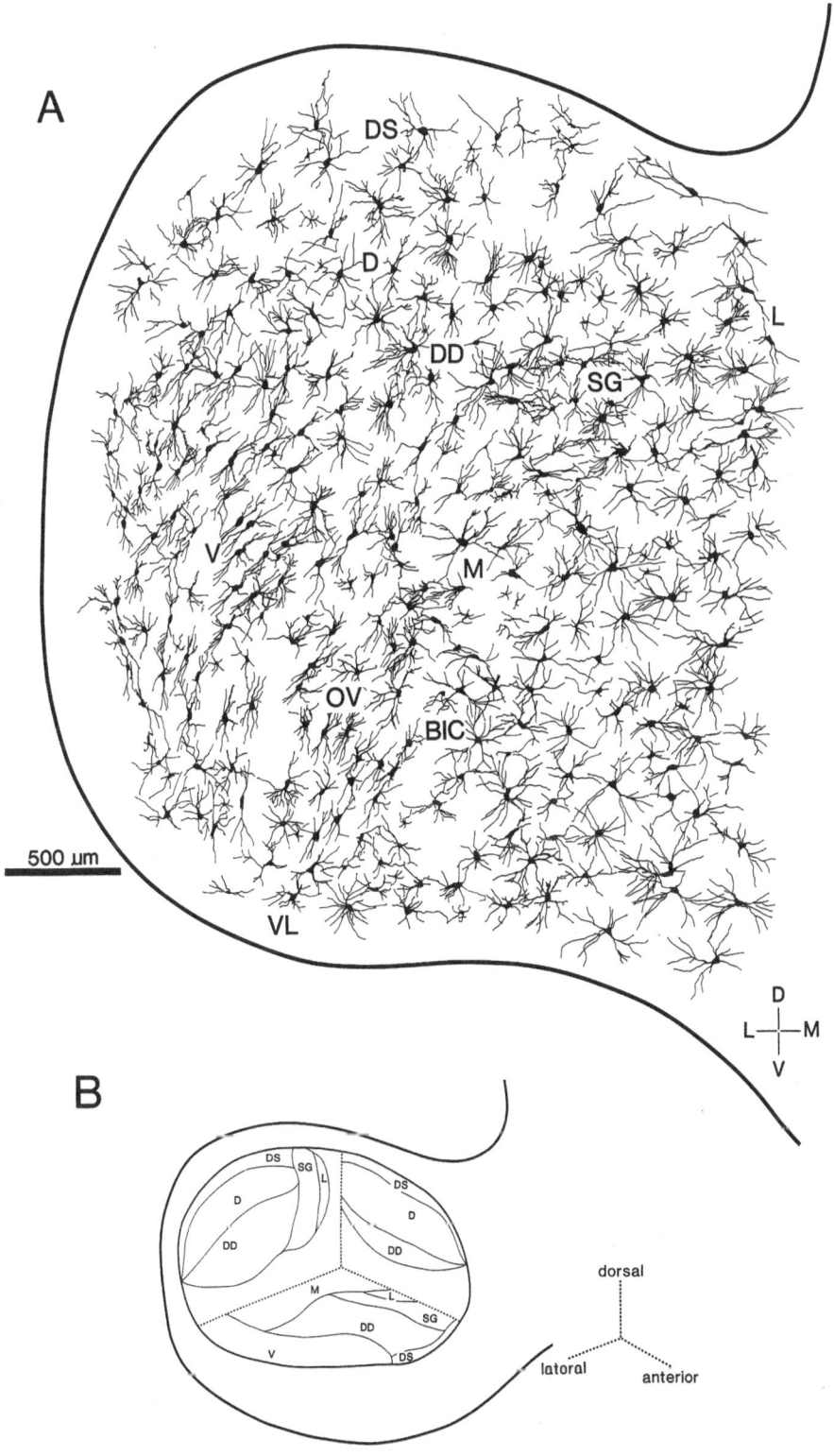

A

DS

D

DD

SG

L

V

M

OV

BIC

500 µm

VL

D
L —— M
V

B

DS SG L
D
DD

DS
D
DD

M
L SG
DD
V DS

dorsal

lateral   anterior

5

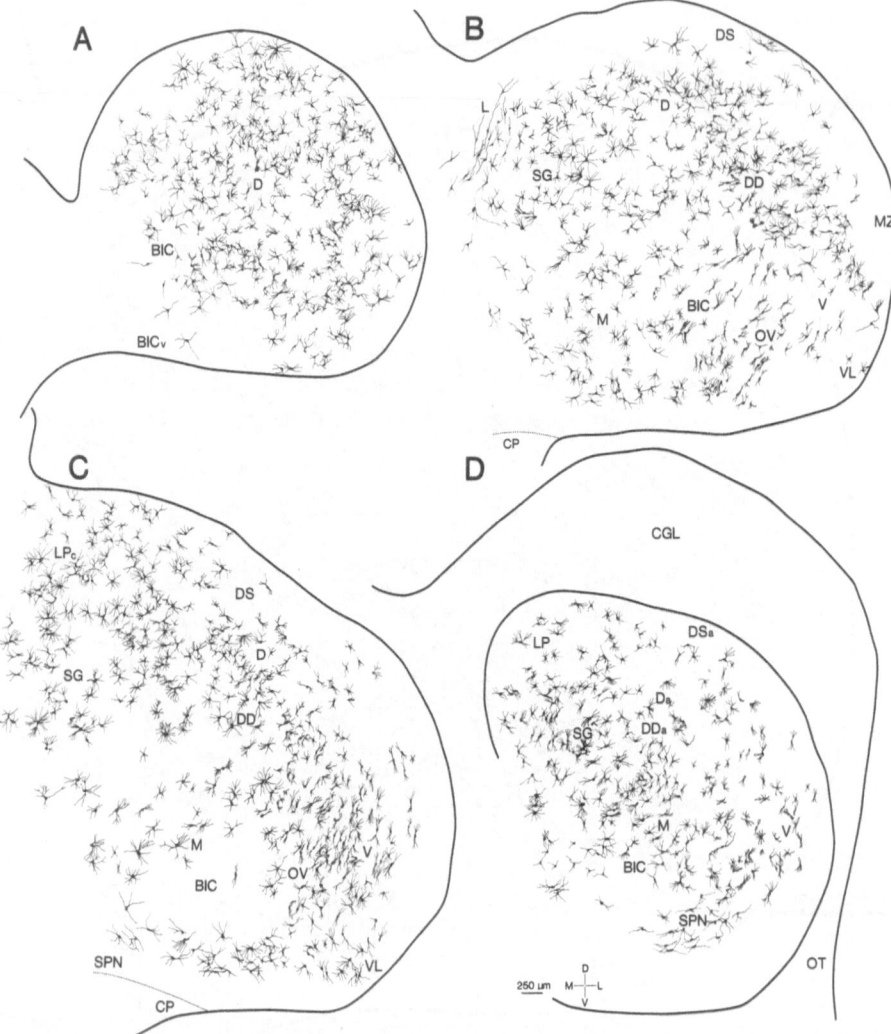

Fig. 2 A–D. Golgi reconstructions of the neuronal architecture at four transverse levels of the medial geniculate body. **A** A section 400 μm from the posterior extremity and consisting entirely of the caudal part of the dorsal nucleus. **B** A section 1600 μm from the caudal pole; note that the posterior limitans nucleus is prominent only at this level. **C** A section 2800 μm from the caudal extremity; note the areas of cellular rarefaction through which fibers pass. **D** A section 3500 μm rostral to the caudal pole; most nuclei here are reduced in size, the anterior nuclei of the dorsal division are present, and the bushy cells from the rostral part of the medial division are conspicuous (above *M*). Each reconstruction was made by superimposing two serial, 100-μm-thick sections. (From Winer and Morest 1983a) Golgi-Cox method, adult cat; planachromat, N.A. 0.18, × 75

preparations. The *dorsal division* (Fig. 4A, D–H; Fig. 44C) contains some of the smallest cells in the medial geniculate body (Morest 1964; Winer and Morest 1983a), although certain subdivisions (Fig. 4G: suprageniculate nucleus) are exceptional for their large neurons. A weak laminar arrangement of some principal cell dendrites and portions of the ascending axonal plexus is present, though less fully expressed than in the ventral division. Dorsal division cells are much more dispersed than those in the ventral division and form small, irregular clusters. *Medial division* cells are still more scattered (Figs. 3, 4I) and are larger (Fig. 44E) than neurons in the other subdivisions. Numerous medium-sized cells are present here, sometimes in clusters, and the nucleus is punctuated by the many axons passing through it and imposing a reticulated texture on the neuropil. Many of the neurons in each of these thalamic divisions project to the auditory cortex (Winer et al. 1977), whose subdivisions are shown in Fig. 5.

3.3 Neuronal Architecture of the Ventral Division

The ventral division contains the fewest and best-studied cell types in the medial geniculate body. Its stereotyped structure has been described by a number of workers (Ramón y Cajal 1911; Rose and Woolsey 1949; Morest 1964, 1965a). The ventral division occupies most of the ventrolateral quadrant of the medial geniculate body but is not present in the caudal one-quarter (Figs. 1B, 2A, 3). Rostral to this it becomes increasingly prominent (Figs. 1A, 2B–D, 3C, E, G) and constitutes somewhat less than a third of the volume of the medial geniculate body. The ventral division in the present account consists of three architectonically distinct nuclei (see Table 1): the ventral nucleus (*pars lateralis*), the ovoid nucleus (*pars ovoidea*), and the marginal zone. The ventrolateral nucleus is considered as part of the dorsal division.

In Nissl (Figs. 3, 4A, C) and Golgi (Fig. 2) preparations the *ventral nucleus* is conspicuous for the rigid, laminar organization of the neurons, particularly the large thalamocortical relay cells, and for its paucity of different cell types. Their dendrites, and the axons ascending from the midbrain, form regular, laminar arrays (see Sect. 3.4). This relay neuron and a smaller cell with a locally arborizing axon are the only cell types in the ventral nucleus (Morest 1975a). Golgi preparations show (Fig. 6) that the principal cell dendrites are arranged in long, slightly convex rows or laminae (Fig. 1A), which are continuous and contiguous throughout the ventral nucleus (Fig. 2B–D). These laminae are longest in the lateral part of the medial geniculate body and become progressively shortened and curved more medially, where the transitional zone between the ventral nucleus and the ovoid nucleus marks the tortuously curved laminae characteristic of the latter. The laminae are clearest in the caudal two-thirds of the medial geniculate body (Fig. 2B, C) and less obvious in the rostral pole (Fig. 2D) where the fiber architecture tends to obscure them.

The dendrites of the ventral nucleus *principal neuron* emerge from both ends of the elongated, medium-sized soma and branch repeatedly and obliquely to form dendritic bushes or tufts (Fig. 6: *3, 5, 8, 22, 32, 43, 49*). The bushy cells are the most numerous cells in the ventral division (Morest 1964, 1975a) and are apparent, from the size of their soma, in Nissl (Figs. 4A, 8A, C) and Golgi (Fig. 6) preparations. In rapid Golgi impregnations they have complex dendritic arbors (Fig. 7). The long axis of the arbor usually parallels the dorso-ventral plane and is shortest latero-medially. Many secondary and tertiary dendrities run orthogonal to the long axis of the laminae and invade the interlaminar neuropil. From serial reconstructions of Golgi-impregnated sections, the thick-

7

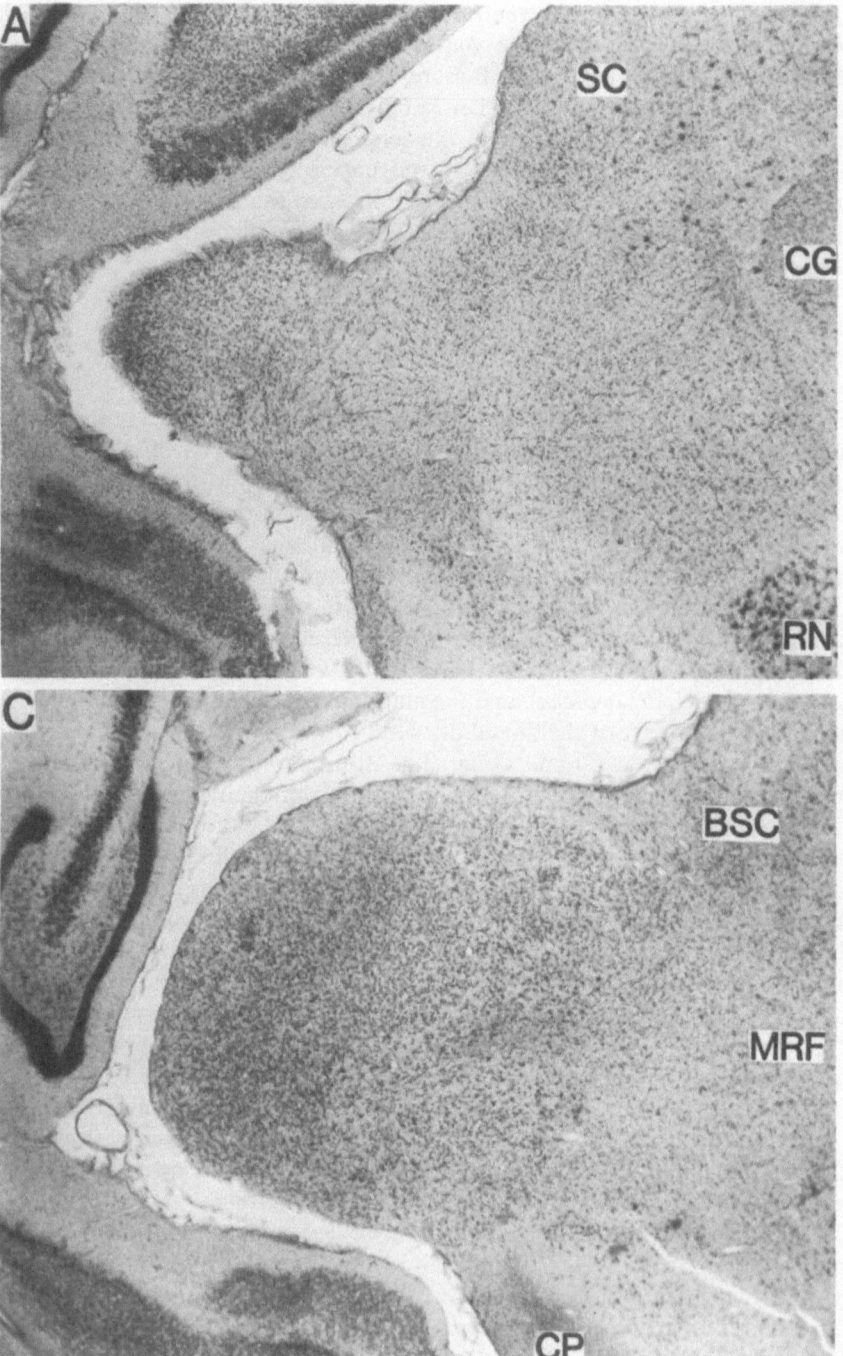

Fig. 3. Photomicrographs (**A, C, E, G**) and outlines (**B, D, F, H**) of transverse, 40-μm-thick, celloidin-embedded sections from the medial geniculate body. **A, B** From the caudal extremity, where only the caudal dorsal nucleus is present. **C, D** At the caudal third. **E, F** At the rostral third; note the large, scattered, and darkly staining cells of the medial division. **G, H** Near the rostral pole (see Fig. 2 D), where the ventral division is reduced. Nissl method, adult cat; semi-apochromat, N.A. 0.04, ×15.6

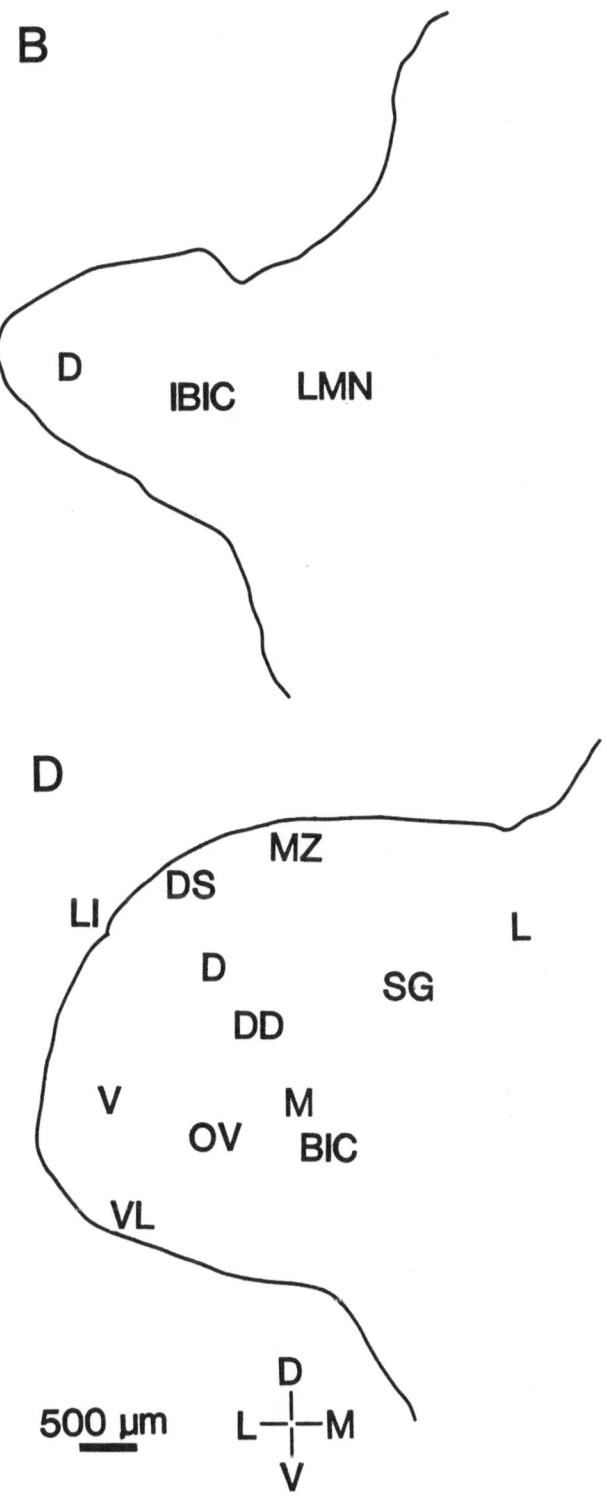

Fig. 3 B, D

9

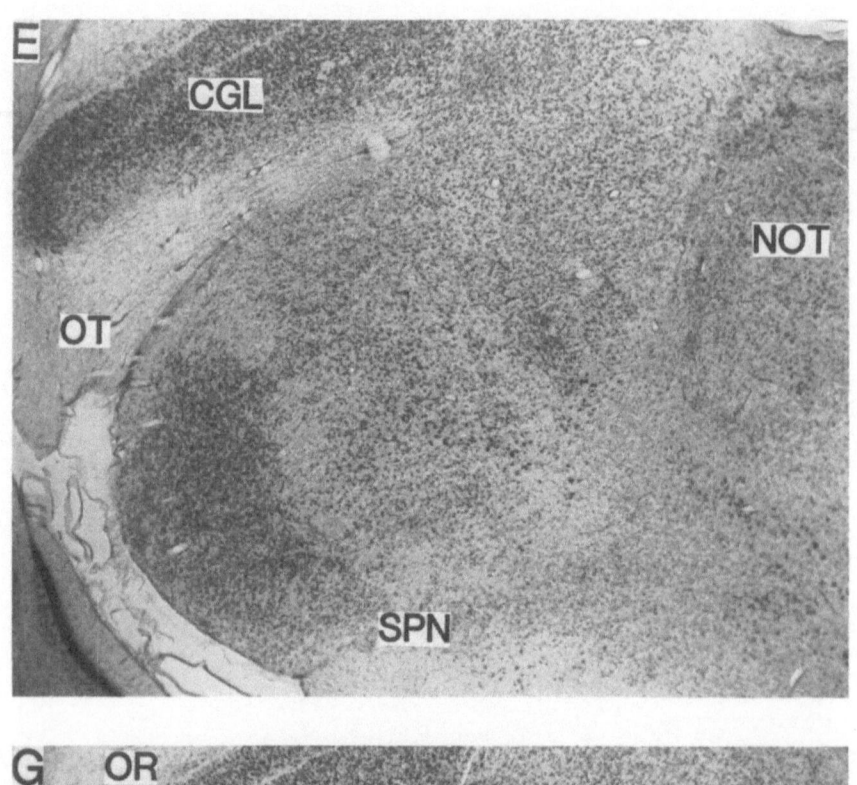

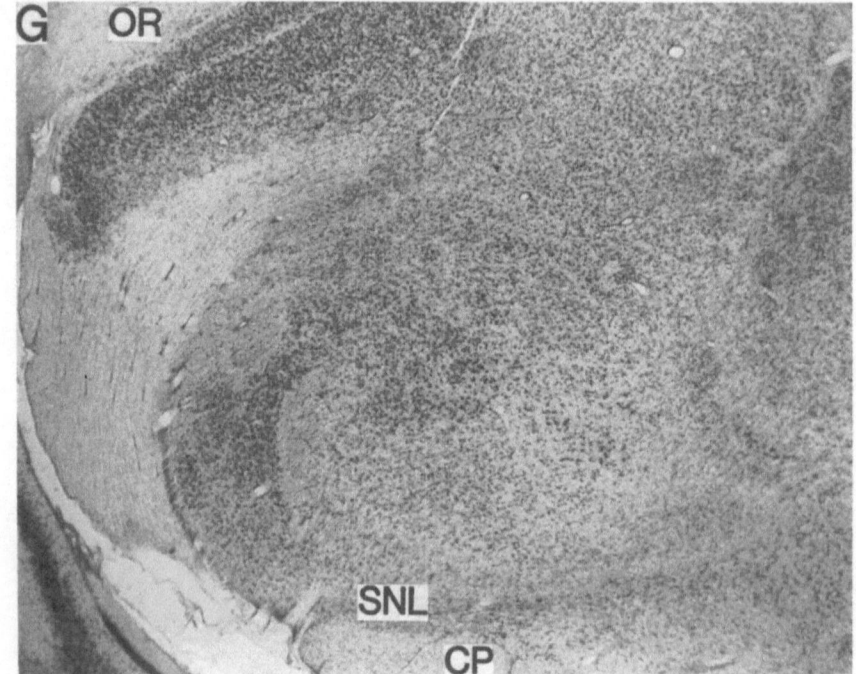

Fig. 3 E, G

10

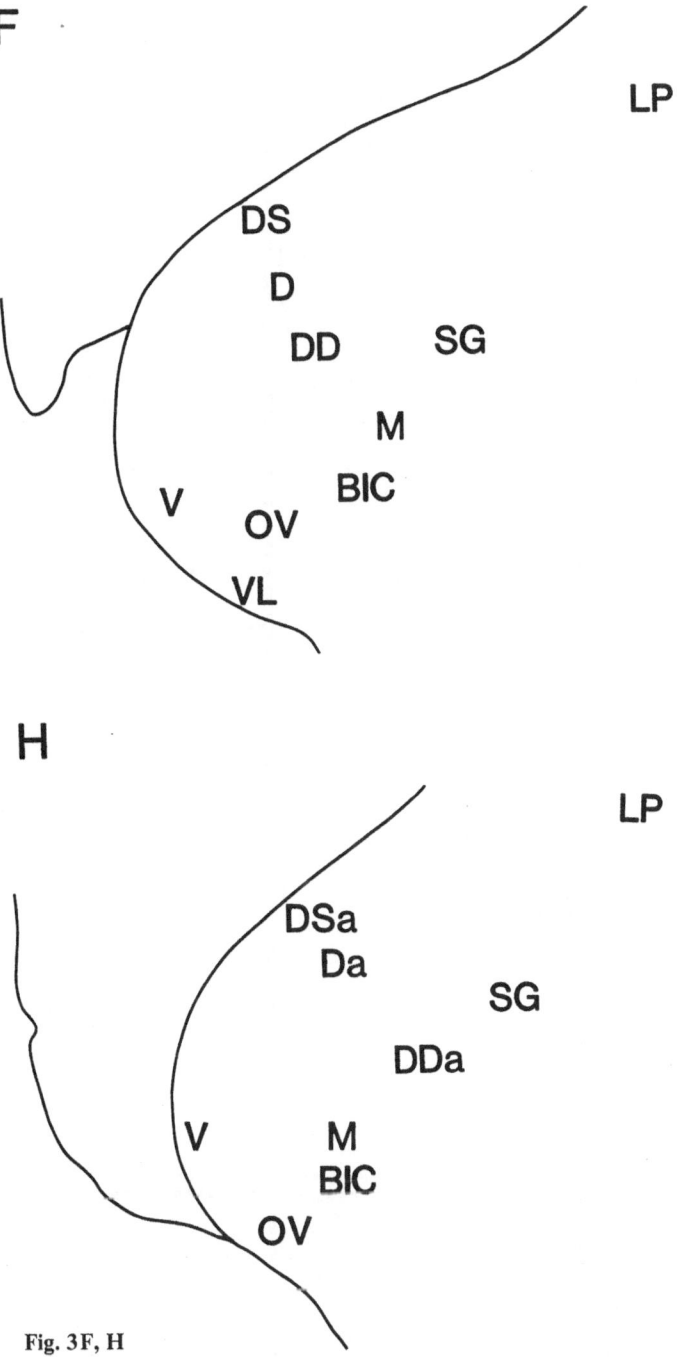

Fig. 3F, H

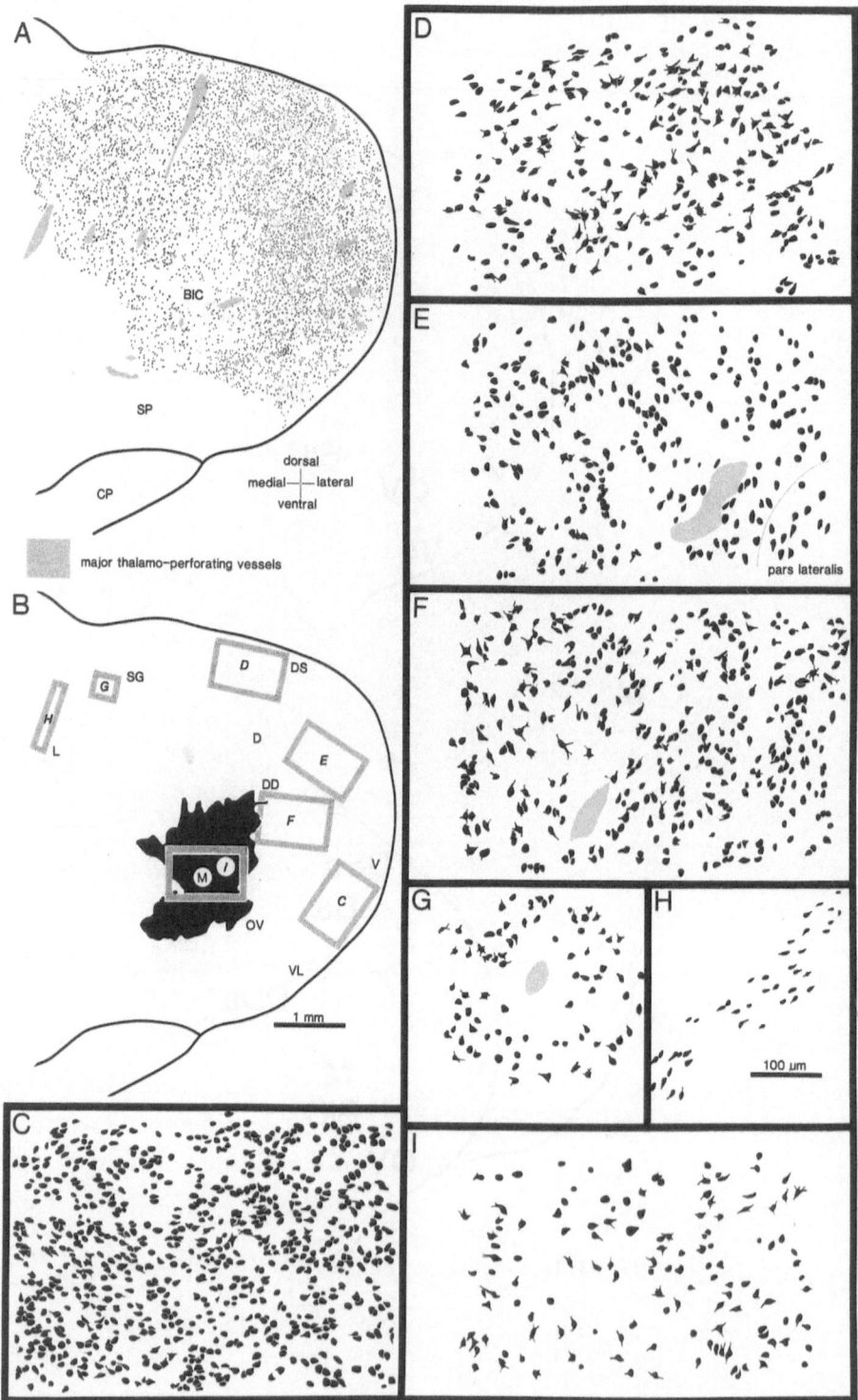

ness of a ventral nucleus fibrodendritic lamina is estimated to be about 50–100 µm (Morest 1965a). As defined here, a single lamina consists of the main part of the dendritic arbor of adjacent principal cells, the adjoining afferent axonal plexus, and portions of intrinsically branching axonal profiles, each regularly arranged with respect to one another and across which an orderly progression of best frequency occurs. The definitive feature is the width of principal cell dendritic domains, since each of the other dimensions is subsumed within this. Since the width of the overlapping principal cell arbors is greater than 50–100 µm, many distal dendrites must cross laminae, and two-to-six principal cell somata could be contained in the short axis.

The dendrites of mature bushy cells are irregularly covered with appendages (Fig. 7) which sometimes extend (though sparsely) to the soma (Fig. 8F). These appendages are heterogeneous in form: some are long and sessile, or long and pedunculated, or short and thornlike; and others have twiglike bifurcations or an intermediate form. The trunk and terminal dendrites are usually free of appendages, although the shaft often is irregular or crenated. Appendages concentrate along, though are not confined to, the intermediate dendrites. The axon of the tufted cell (Fig. 6: *45, 52*; Fig. 7) is about 2 µm thick and usually arises from the soma or, less often, from a dendritic trunk. It turns first dorsally and then medially to join the auditory radiations, although it can rarely be followed this far even in immature animals since it usually becomes myelinated within 30 µm of its origin. These axons are apparently without collaterals.

A *Golgi type II cell* is also present in the ventral nucleus and strongly contrasts with the bushy neuron (Fig. 6: *1, 13, 19, 26, 33, 34*). In Nissl preparations these interneurons are readily identified by their small size, smooth soma, small nucleus, and pale cytoplasm (Figs. 4A, C; 8B–D; 44B). From its round, oval, or drumstick-shaped soma issue four-to-six main dendrites. These slender trunks emerge irregularly from the surface and sometimes give a stellate appearance to the cell (Fig. 6: *44*). As a rule the dendrites divide once or twice — sometimes acutely (Fig. 9) — and often have a tortuous course. They project toward the dendrites of adjacent principal cells, among which they branch. Smooth den-

◁ **Fig. 4A–I.** The contrasting cytoarchitecture of the subdivisions of the medial geniculate body. Glial cells have been excluded from this figure. **A** Cell-stained, 15-µm-thick, paraffin-embedded section to show regional architectonic variations at low power. Nissl method, adult cat; planachromat, N.A. 0.2, ×48. **B** The locations from which the neurons in C–I were drawn. The brachium of the inferior colliculus is shown also (*solid black*). **C** Neurons from the ventral nucleus; in this panel, *dorsal* is to the *left,* and *lateral* is the *upper edge.* Note the high packing density and preferential orientation of these cells. For C–I: planachromat, N.A. 0.65, ×500. **D** The cytoarchitecture of the superficial dorsal nucleus is characterized by round or triangular somata and numerous small perikarya. **E** The cytoarchitecture of the dorsal nucleus, where medium and small cell bodies predominate, and in which cell-poor islands of neuropil occur. Where this section encroaches on the dorsomedial part of the ventral division (*lower right*) there is an abrupt transition in structure. **F** The neurons of the deep dorsal nucleus, where multipolar radiate neurons and small cells are common and the packing density is somewhat greater than in other parts of the dorsal division. Where these neurons abut the medial division, along the ventromedial border, larger cells occur. **G** Cytoarchitecture of the suprageniculate nucleus showing its scattered, large and small cells and their low packing density. **H** The medium-sized and small neurons of the posterior limitans nucleus and their characteristic dorsolateral-to-ventromedial orientation. **I** The heterogeneous perikarya of the medial division; the low packing density, lack of obvious orientation, and somewhat larger somata are clear

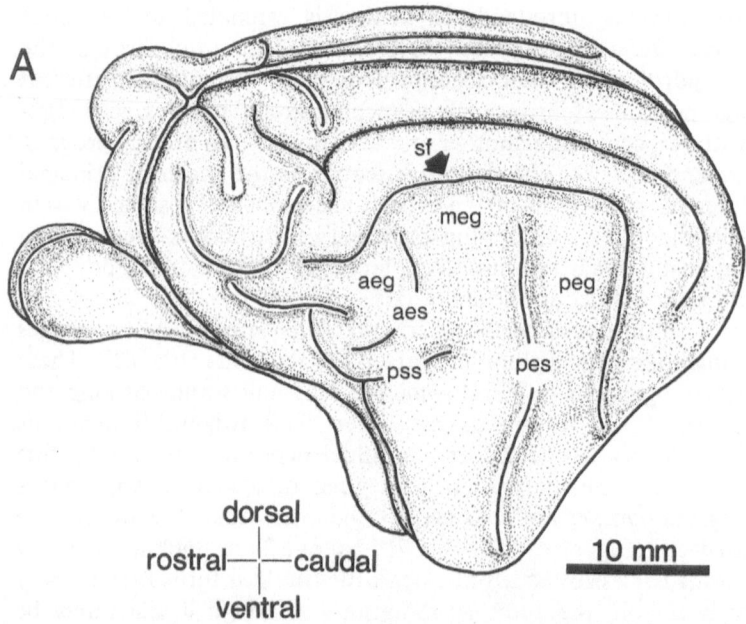

A

sf

meg

aeg

aes

peg

pss

pes

dorsal

rostral——caudal

ventral

10 mm

Fig. 5 A–C. Subdivisions of cat auditory cortex. **A** Major gyri and sulci of cat cerebrum. **B** Schematic view of cytoarchitectonic and electrophysiologically defined subdivisions based on the work of Rose (1949), Rose and Woolsey (1949, 1958), Woolsey (1961), and Merzenich et al. (1975). **C** Distribution of tonotopic maps and auditory cortical areas from Reale and Imig (1980). The buried, sulcal cortex is unfolded (*hatched areas*).The gradient of *low*-to-*high* frequency is also shown

dritic segments alternate with small clusters of appendages, many of which are longer, more delicate, and less regular than those of principal cells. Many spines are pedunculated; others impose a dentate or sawtooth profile to the dendritic surface. Besides the smaller, flask-shaped soma and unique dendritic form of the Golgi type II cell, its axonal structure distinguishes it from the tufted neuron. The 1- to 1.5-μm-thick axon branches three-to-ten times, and each collateral curves sinuously as it divides. The target of this axon, besides the sparse system of recurrent collaterals to its dendrites, appears to be the dendrites of nearby principal cells. Most collaterals appear to terminate within about 300 μm of their origin (Fig. 9), but some branches may leave even very thick sections. Within one section, several Golgi type II cells may project to a single principal cell to form chainlike arrangements, and individual interneurons may project to several principal cells some distance away (Morest 1975b). No cells without an axon have been observed.

The cells of the *ovoid nucleus* closely resemble neurons in the lateral part of the ventral nucleus. In both Nissl (Figs. 3, 4A) and Golgi (Figs. 1A, 2B–D) preparations, however, this nucleus is distinct from the ventral nucleus. Morest (1965a: Figs. 1, 2) described ovoid nucleus laminae as compact and coiling about each other in short, irregular rows which impart a tortuous pattern, in contrast with the smooth and gentle arcs described by ventral nucleus laminae. Second, the fibrodendritic laminae of the ovoid nucleus, which abut the medial

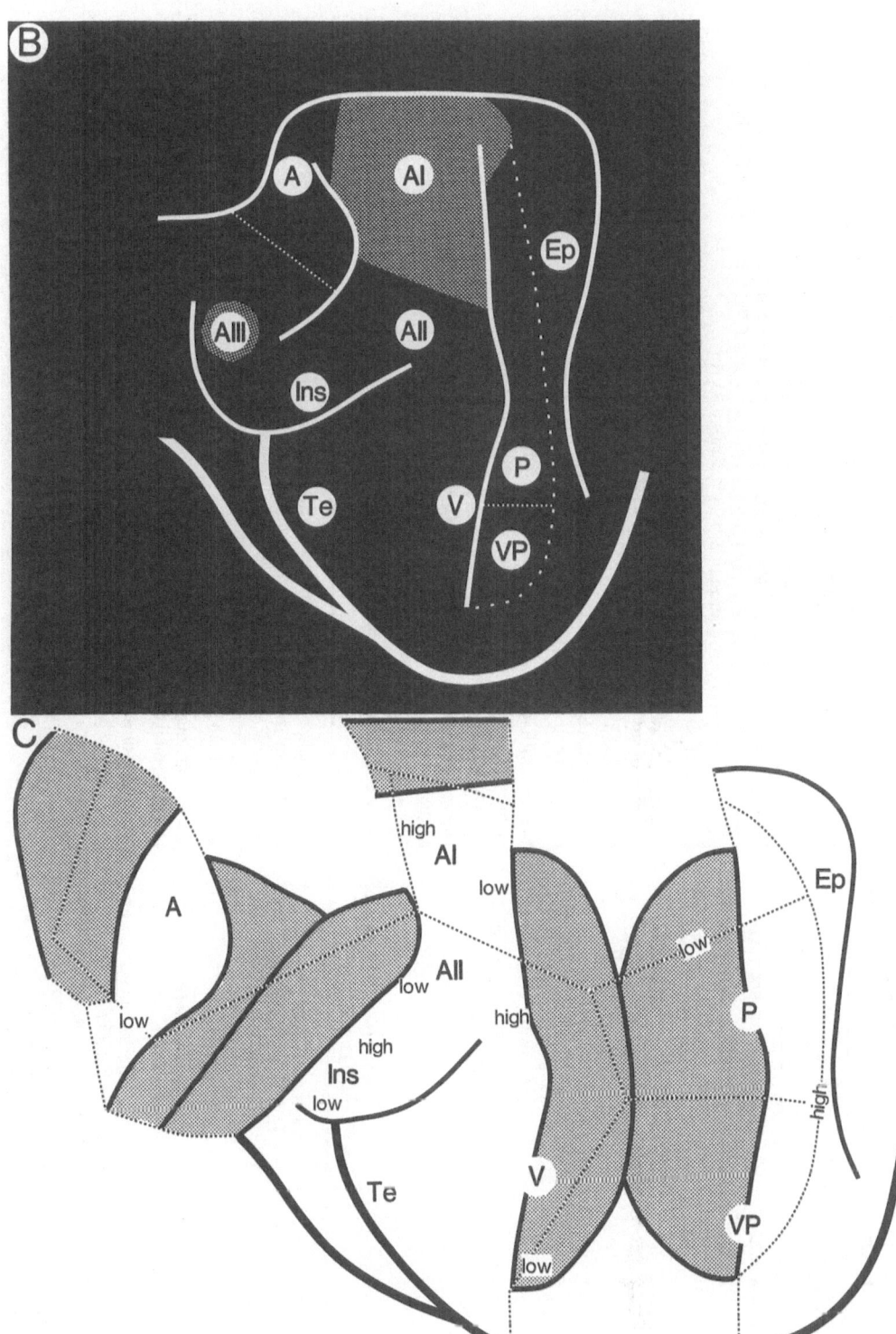

Fig. 5 B, C

Table 1. Architectonic subdivisions of the medial geniculate body

Reference	Division and/or nucleus	Description of cell types and distinctive features	Species	Staining methods
Ramón y Cajal (1911)	1. Dorsal (or superior) nucleus	18–24 μm in diameter polygonal or triangular neurons, sparse Nissl substance (…un protoplasma peu fourni en amas chromatique…, p 285); fusiform or star-shaped cells predominant with multiple, thin, radiating, dendrites	Cat, dog, mouse, guinea pig, rabbit	Nissl, rapid Golgi, Weigert-Pal
		Neurons of small dimensions, scattered without order, and having short axons		
		Recognizable by abundance and extent of intercellular axo–protoplasmic plexus		
	2. Inferior (or ventral) nucleus (including *nucleus ovoidea*)	Neurons with long axons: star-shaped cells whose dendrites have tufts of characteristic filaments; axon may have collaterals		
		Neurons with short axons: ovoid, fusiform, or triangular shape with 3–4 dendrites; locally arborizing axon with varicose appendages		
	3. Medial (deep or magnocellular) nucleus	Large multipolar triangular fusiform or pyramidal cells		
Rioch (1929); Rose and Woolsey (1949)	1. *Pars principalis*	Closely packed, small neurons; cells lying dorsally in horizontal sections are somewhat larger, and less densely arranged; in lateral-to-medial sections there is a decrease in the cell population and an increase in cell size	Dog, cat	Nissl
	2. *Pars magnocellularis*	Loosely arranged, large cells with large, pale nuclei, and small cells with deeply stained nuclei		
	3. Posterior nuclear group (includes posterior nucleus, suprageniculate nucleus, and possibly parts of pulvinar)	No description given		

			Cat, opossum, bat, chick, raccoon, gerbil, squirrel, macaque monkey, baboon	Rapid Golgi, Golgi-Cox, Golgi-Kopsch, Nissl, Weil, Weigert, Richardson, Bodian, Klüver-Barrera
Morest (1964); present account	Ventral division			
	1. *Pars lateralis*	Cells with tufted dendrites and whose axons project outside the medial geniculate body		
		Neurons with a short axon		
	2. *Pars ovoidea*	Resembles *pars lateralis* except for coiled laminae		
	3. Marginal zone	Largely cell-free zone of neuropil on the free surface		
Morest (1964); Winer and Morest (1978, 1983a); present account	Dorsal division		Same as above	Same as above
	1. Caudal dorsal nucleus	Principal cells with stellate, evenly spaced dendrites		
		Principal neurons with bushy dendrites		
		Small, stellate neurons with locally arborizing axon		
		Larger neurons with locally arborizing axon		
	2. Superficial dorsal, dorsal, and deep dorsal nuclei; and anterior superficial, dorsal, and deep dorsal nuclei	Same as (1) except axonal plexus attributable to Golgi type II axons is well developed; neurons with bushy or flattened dendrites prominent in the superficial dorsal nucleus		
	3. Suprageniculate nucleus	Large stellate neuron with 6–10 primary dendrites		
		Small flask-shaped or stellate cell with a locally arborizing axon		
	4. Posterior limitans nucleus	Medium-sized principal cell with sparse polarized and elongated dendrites		
		Small neuron with stellate dendrites and a locally arborizing axon		
	5. Ventrolateral nucleus	Now considered as part of the dorsal division; large principal neuron with 6 primary dendrites and a spherical dendritic field		

Table 1 [continued]

Reference	Division and/or nucleus	Description of cell types and distinctive features	Species	Staining methods
Morest (1964); Winer (1979); Winer and Morest (1983b); present account	Medial division	Medium-sized neuron with bushy dendrites	Same as above	Same as above
		Medium-sized neuron with wide-field or elongated dendrites		
		Medium-sized neuron with radiate dendrites		
		Magnocellular neuron with irregularly stellate dendrites		
		Small stellate neuron with slender dendrites and locally arborizing axon		
Oliver (1982)	1. Ventral division	Principal neuron with bushy dendrites	Tree shrew	Golgi-Cox, rapid Golgi, Golgi-Kopsch, Nissl
		Small cell with thin dendrites, and pedunculated appendages		
	2. Dorsal division	Caudomarginal zone contains cells whose disc-shaped dendritic fields avoid ventral nucleus laminae		
	Dorsal nucleus	Principal cell with ovoid or spherical dendritic field		
	Deep dorsal nucleus	Principal cell with cylindrical dendritic field		
	Anterodorsal nucleus	Principal neuron with spherical dendritic field		
	Suprageniculate nucleus	Large principal cell with ovoid dendritic field		
		Small stellate Golgi type II cell		
	3. Medial division			
	Rostral nucleus	Large round soma; spherical, truncated dendritic field		
		Cell with spindle-shaped soma and larger dendritic field		
	Caudal nucleus	Magnocellular neuron with large round soma; spherical, truncated dendritic field		
		Wide-field neuron with hemispheric dendritic field		

18

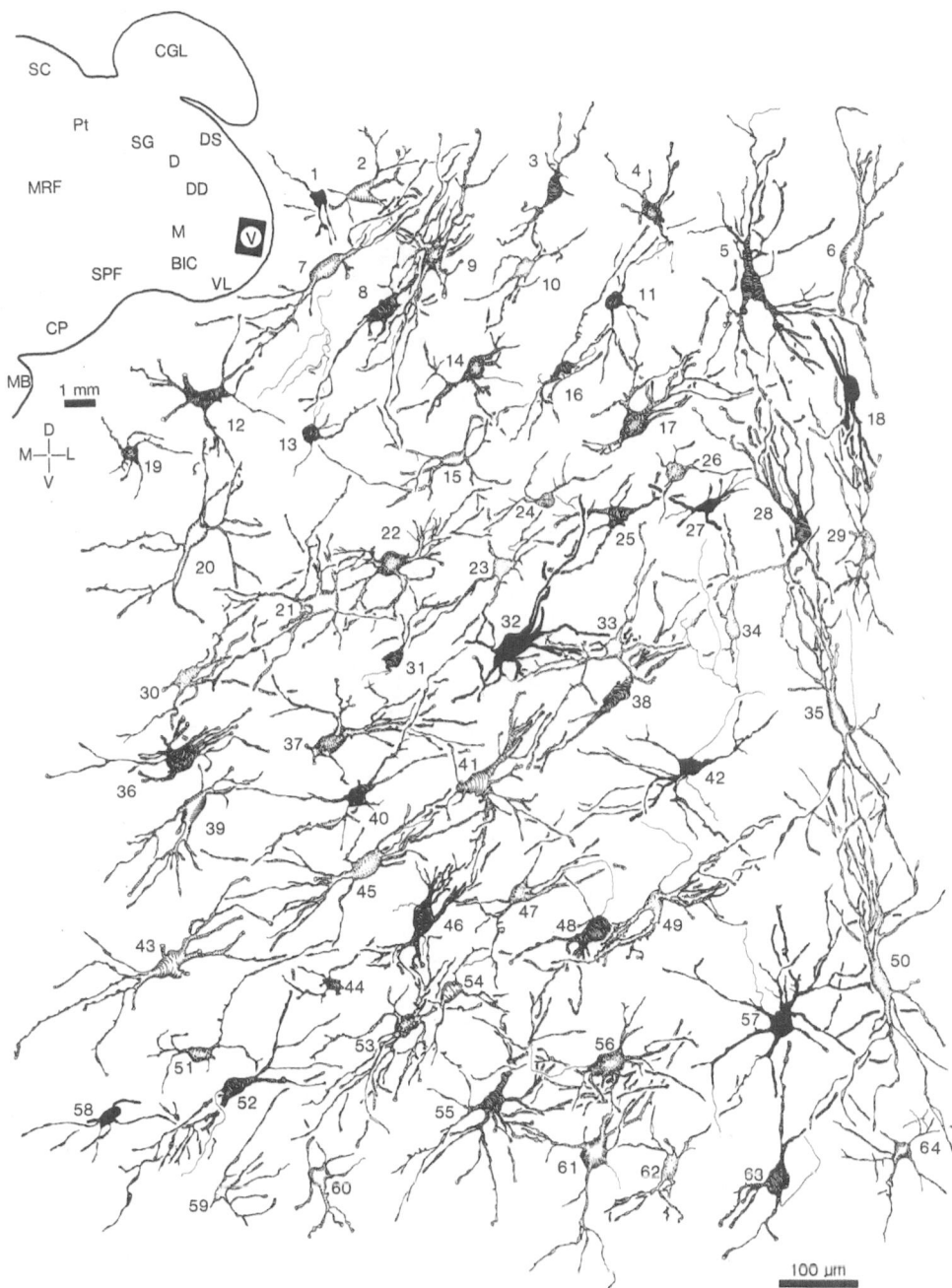

Fig. 6. The varieties of neurons from the ventral nucleus. Two main cell types are present: larger cells with bushy dendrites (e.g., *6, 28, 35, 49*) and smaller flask-shaped neurons with a weakly stellate configuration (e.g., *13, 19, 44, 64*). See text for further explanation. This survey was made by drawing neurons from six serial, 100-μm-thick sections. In this and the following figures, where a process leaves or enters the section is indicated by a hollow profile. Golgi-Cox method, adult cat; semi-apochromat, N.A. 1.05, × 375

19

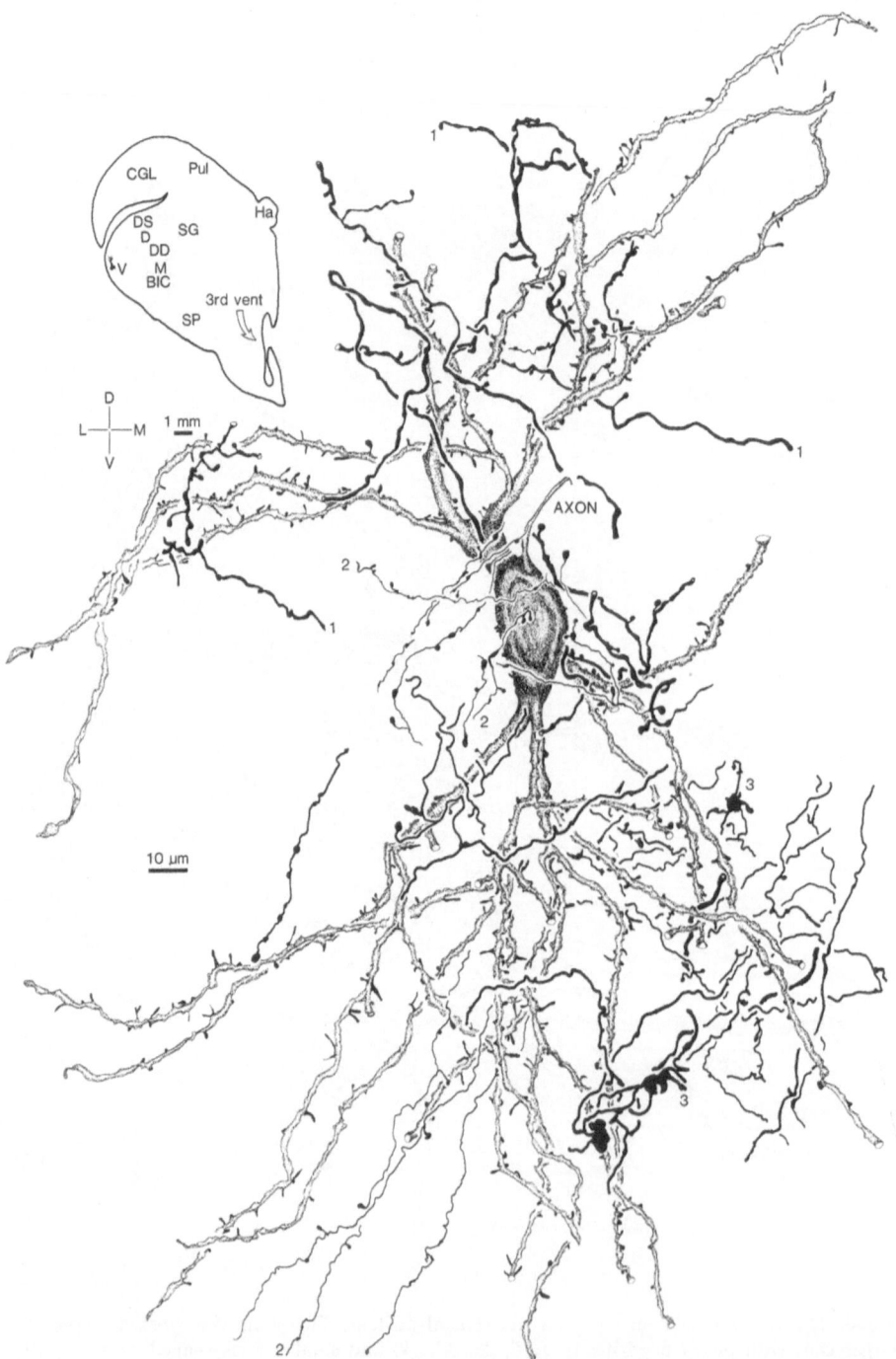

Fig. 7. A typical bushy principal neuron from the ventral nucleus of the medial geniculate body. Note the polarization of the dendrites and their relationship to the plexus of afferent axons. Conspicuous among the latter are the axons of Golgi type II cells (beneath the soma and with several collateral branches) and extrinsic (*1*) thick and (*2*) thin fibers with various knoblike swellings along their length. These fibers are probably of ascending origin (Morest

division, are progressively shortened, obscured, and crossed by the fascicles of axons from the brachium of the inferior colliculus entering the medial geniculate body. The degree to which the laminae in the ovoid nucleus are compressed and truncated as they pass medially is uncertain. These observations are consistent with the variability of best frequencies seen in electrode penetrations through the ovoid nucleus, compared with the ventral nucleus (Morel and Imig 1982). In Golgi preparations similar kinds of cells are present, although their dendritic fields are less regularly aligned than those of comparable neurons in the ventral nucleus. The inclination of the laminae in the medio-lateral plane (as opposed to the predominantly dorso-ventral orientation in the ventral nucleus) is also conspicuous (Fig. 10).

The architecture of the *marginal zone* resembles that in the ventral division in some respects but is different in other ways. In Nissl (Figs. 3, 4A) and Golgi (Fig. 6, *lateral edge*) and fiber-stained (Fig. 45A, B: *arrow*) material it is a slender, crescentic rim on the ventrolateral, free border of the medial geniculate body and about 100 μm thick. It is remarkable for the sparsity of its cells, its pale staining in both cell and most fiber techniques, and the characteristic texture of the neuropil, which is dominated by long, curving, and extremely fine axons. The structure of the principal cells is also unusual: their bushy dendrites are oriented in a rigid, nearly vertical array almost without curvature (Fig. 6: *6, 18, 35, 50*).

The *ventrolateral nucleus* occupies only a small sector wedged between the ventral nucleus and the marginal zone. On the basis of its neuronal architecture and cortical connections, it is considered as part of the dorsal division (see Sect. 3.9).

3.4 Structure of Axons in the Ventral Division

The dendritic laminae dominating the ventral division receive axons primarily from three sources. The ascending fibers impose a characteristic texture on the neuropil, entering the ventral nucleus from the brachium of the inferior colliculus and forming parallel sets of axonal laminae (Fig. 11A) apposed to and dividing along the main axis of the dendritic laminae. Their form and trajectory suggest they are of midbrain origin (Morest 1964, 1965a, b). Many of these fibers form a dense, peridendritic plexus around principal neurons (Fig. 7: *1*) and Golgi type II cell (Fig. 9: *2*) dendrites. These axons are about 2 μm thick and have beadlike swellings at irregular intervals (Fig. 11A: *1*). Their terminals continue for some length within a lamina, occasionally emitting slender collaterals (Fig. 12C: *1, arrowhead*) before ending or passing from the section

◁ 1964, 1975a; Jones and Rockel 1971). Axosomatic endings occur but are less common than axodendritic terminals (*2*). Possibly descending axons, whose targets are probably the ventral and distal dendrites of these cells, are also shown (*thin outlines, lower right*). Some morphologically unusual endings (*3*) with large preterminal swellings are present. The dendritic fields of ventral nucleus neurons are usually wider medio-laterally than those of similar cells in the ovoid nucleus (see Fig. 10). Rapid Golgi method, 41-day-old cat; semi-apochromat, N.A. 1.32, × 1250

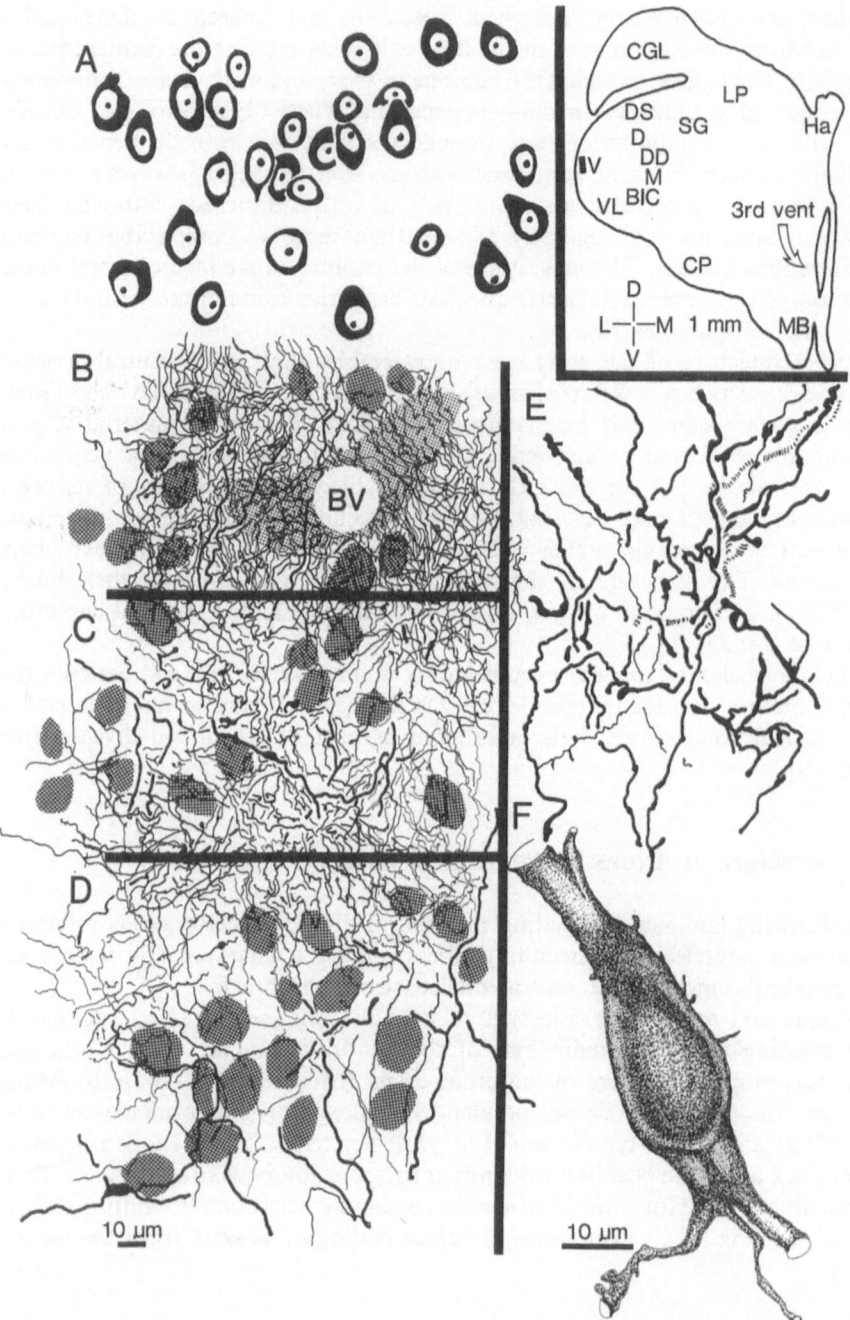

Fig. 8A–F. The relationship between the cytoarchitecture and axonal architecture in the ventral division. **A** The medium-sized, elongated principal neuron somata (see also Fig. 4C) and the small, oval, or flask-shaped perikarya are distinct in a 15-μm-thick, paraffin-embedded, Nissl-stained section uncorrected for shrinkage and from an adult cat. Semi-apochromat, N.A. 1.25, ×787.5. **B** The texture of the neuropil in the ventral nucleus when all impregnated axons are drawn. A dorsal-to-ventral orientation is evident, and the only areas of rarefaction are near somata (*hatched outlines*). For **B–D**: rapid Golgi method, 41-day-old cat; semi-apoch-

(Fig. 12 C: *1, three-sided arrowhead*). Sometimes these axons extend far enough medio-laterally to cross the borders of laminae (Fig. 12 C: *1, 3*), possibly to influence adjacent laminae. The size of these midbrain axons, particularly their terminal fields, may be larger than previous reports suggest (Morest 1964). In the present study they occupy slabs which extend at least the width of a lamina (Fig. 12 A: *1*). Their distribution along the length and depth of a lamina cannot be ascertained due to the curvature of the lamina. Their telodendria often have slender, budlike collaterals extending from the main branch, and delicate *boutons terminaux* (Fig. 12 A: *2*). Their collaterals may form perisomatic endings on principal (Fig. 12 A: *3*) and Golgi type II (Fig. 12 A: *4*) perikarya as well as dendrites (Fig. 12 A: *1*). The translaminar component of these axons is not as conspicuous in Golgi preparations as the intralaminar plexus (compare Figs. 11 A: *2* and 11 B, C); in single sections where every impregnated axon is drawn (Fig. 8 B) the translaminar axons are obscured. When the laminar, vertical contribution of midbrain axons forming slablike terminals (Fig. 12 A) is omitted, the overlap of this system with the translaminar endings (Fig. 8 B, C) is evident. As noted above, these (and other) axons sometimes make perisomatic endings near principal (Figs. 7, 8 E, F) and Golgi type II cells (Fig. 9). Similar patterns of axonal organization appear to prevail in the ovoid nucleus (Fig. 10: *1, 2*), although the complexity of the axonal plexus and its refractoriness to impregnation make it difficult to be certain.

A second group of extrinsic axons to the ventral division is probably of cortical origin, since profiles similar to these degenerate after large lesions to the auditory cortex (Morest 1975a). These fibers may be thin and beaded (Fig. 7: *2*) or thicker and beaded (Fig. 8 E, *upper right*; Fig. 12 C: *5*). They cross the laminae obliquely and have dorsolateral (Fig. 12 B: *3*) or dorsomedial (Fig. 12 C: *4*) trajectories. The thin endings (Fig. 7: *2*) are often less than 1 μm thick, while the larger axons (Fig. 12 B: *1*) may be 2–3 μm in diameter. The swellings are less frequent and larger on these axons than those of midbrain origin. These endings have fork-shaped collaterals, usually at regular intervals, and follow a relatively direct path through the ventral division. A third component of the ventral division axonal plexus is the collateral system of Golgi type II axons (Fig. 9). These are common near both principal (Fig. 7, *upper left*) and Golgi type II (Fig. 9) cells and contribute heavily to the axonal plexus with their very fine collateral system (Fig. 8 B–E). Their distinctive local course, sinuous and abrupt trajectory, and characteristic smoothness distinguish them from the axons described above.

Besides these three groups of axons, at least one other type of axon, hitherto undescribed and of unknown provenance, has been seen (Fig. 11 D). This large, serpentine ending has more collaterals and is confined to a smaller area than Golgi type II axons, and its morphology is distinctive. On entering the ventral nucleus it gives off many collaterals which form elaborate nests limited to small

◁ romat, N.A. 1.25, × 787.5 **C** Some horizontally oriented, fine axons in the neuropil of the ventral nucleus. **D** The plexus of ascending afferent axons; there are roughly equal intervals between the groups that form laminae (see also Fig. 10). For **E, F**: rapid Golgi method, 41-day-old cat; semi-apochromat, N.A. 1.32, × 1250. **E** The terminal plexus of various extrinsic axons at higher magnification. **F** A principal neuron with somatic appendages

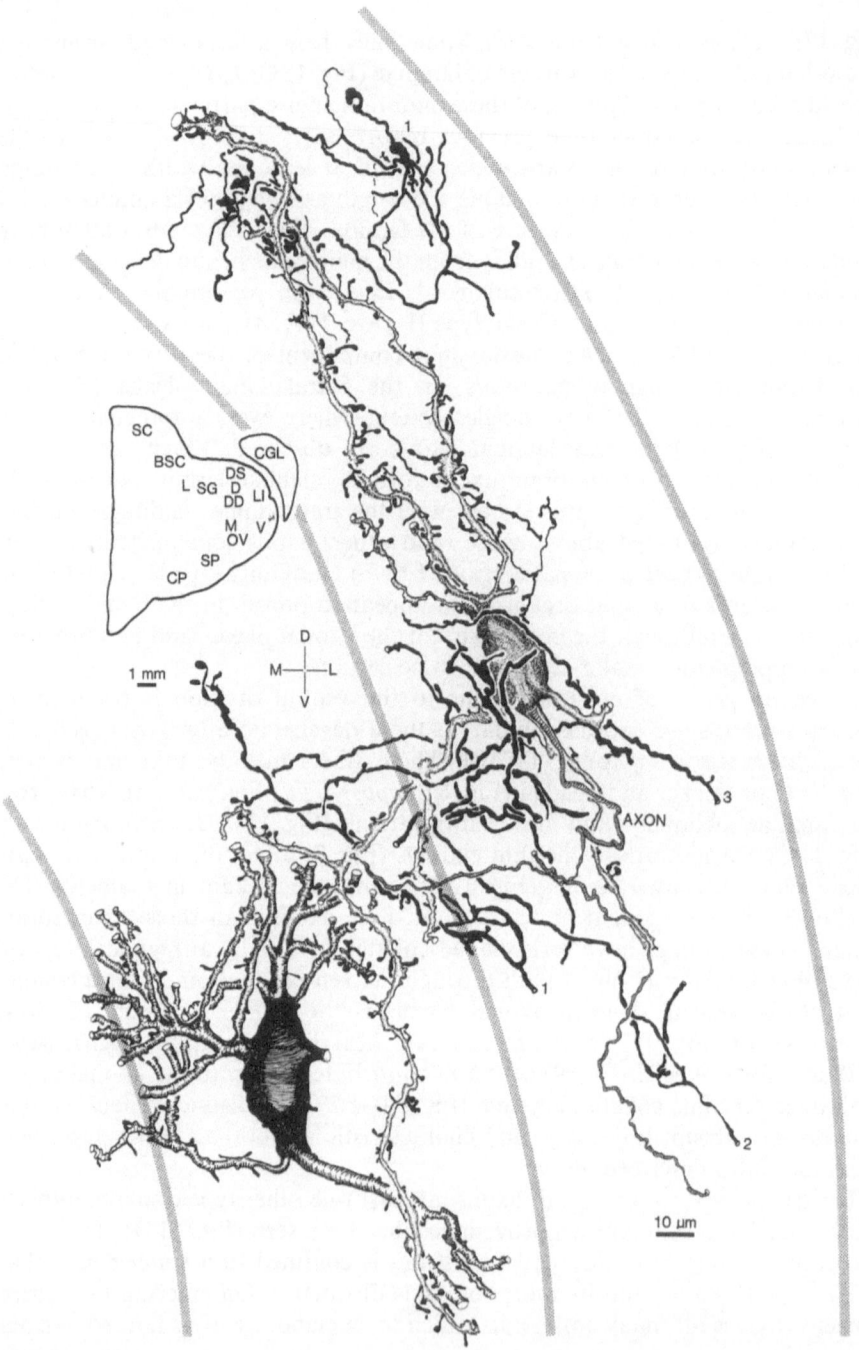

Fig. 9. A Golgi type II cell from the ventral nucleus and its relationship to principal cell dendrites. The processes of the Golgi type II cell (*stippled*) are confined primarily to a single laminae (whose boundaries are shown by *fine stippling*), and its dendrites commingle with those of the principal cell (*hatched*). The dendritic appendages of local circuit neurons are fewer, thinner, longer, and more elaborate than those of principal cells (see Fig. 7). A portion of the axonal plexus (*1–3*) is also shown. The internuncial cell's axon is confined to a few laminae and ramifies among the principal neuron's dendrites and those of adjacent cells. Rapid Golgi method, 41-day-old cat; semi-apochromat, N.A. 1.32, ×1250

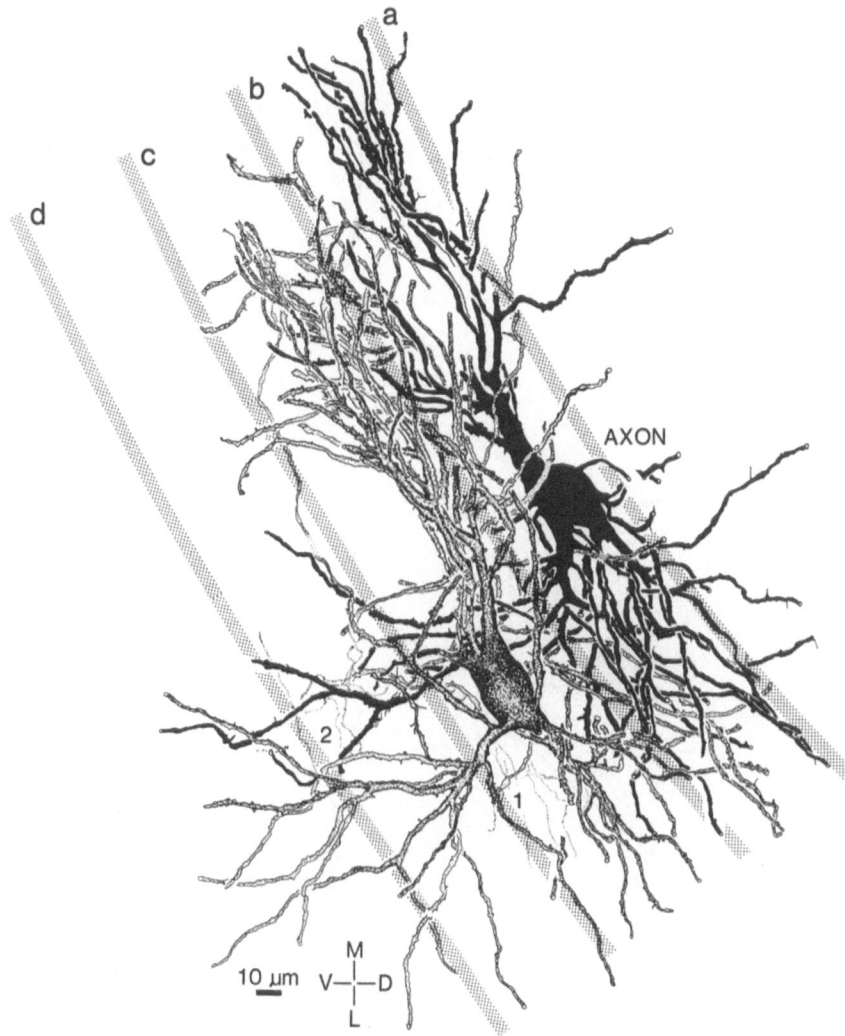

Fig. 10. Principal neurons with bushy dendrites from the ovoid nucleus of the ventral division. These cells have rigidly aligned, parallel dendritic arbors which conform with and contribute to the coiled fibrodendritic laminae of the ovoid nucleus (Morest 1964). These laminae are shorter and more curved than those in the ventral nucleus. The approximate borders of laminae are also indicated (*a–d*). Despite the predominantly parallel arrays of dendrites, some dendritic segments of each cell have translaminar arbors. Axons afferent to proximal (*1*) and intermediate (*2*) dendrites. (From Morest and Winer to be published) Rapid Golgi method, 39-day-old cat; planachromat, N.A. 1.32, ×1250

areas. Some branches project to the marginal zone (Fig. 11 D: *1*) but most are confined to the lateral part of the ventral nucleus. It is unlikely that these endings represent the differentiating processes of immature midbrain terminals, for their collateral system is more extensive than midbrain axons and their shape is different. Some morphological features resemble those of growing axons (Morest 1969), such as growth cones (Fig. 11 D: *1, upper right*). That these

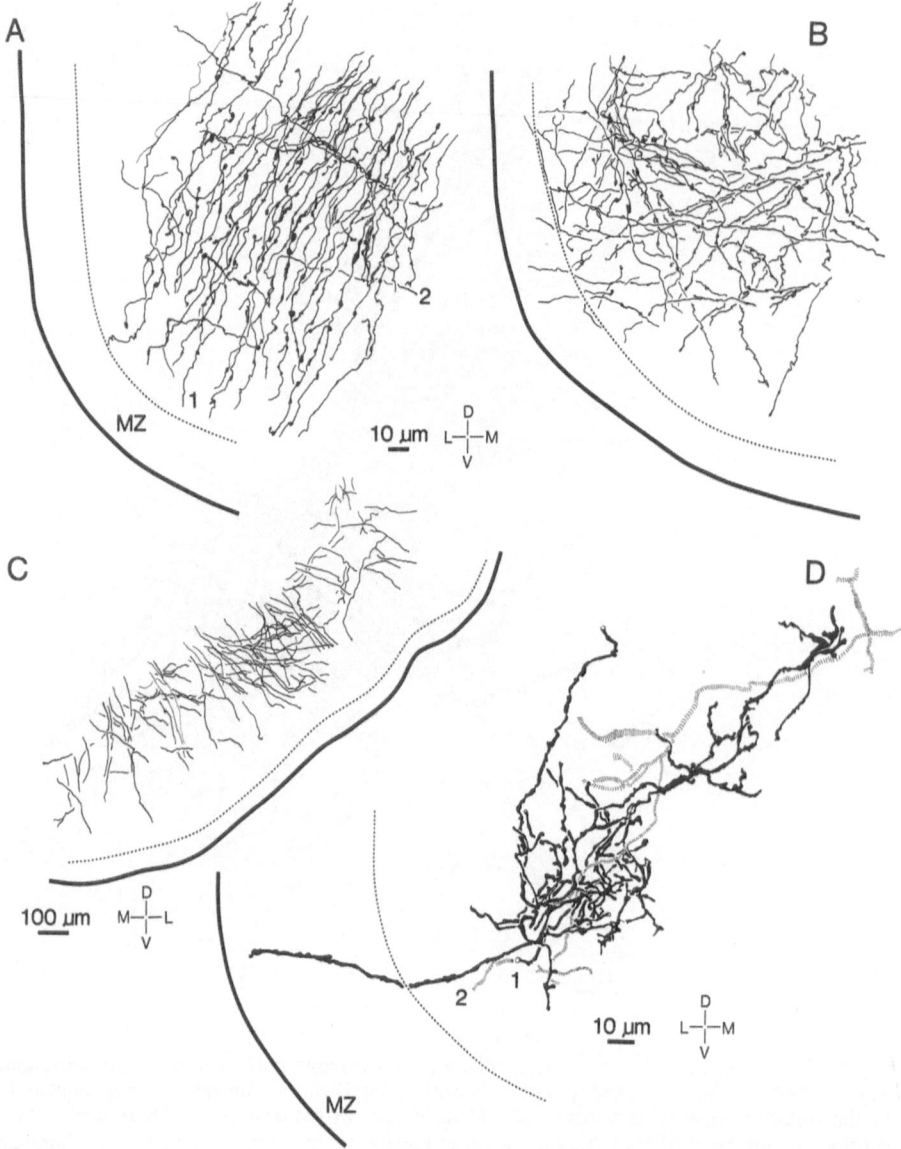

Fig. 11 A–D. Variations in the plexus of axons afferent to the ventral nucleus. **A** The ascending brachial axons and a part of their terminal plexus (see also Morest 1964, 1965a). Besides these dorsally directed axons, which comprise the bulk of the laminae (*1*), some horizontally coursing axons are also present (*2*). For **A, B:** rapid Golgi method; planachromat, N.A. 1.32, ×1250. **B** Some horizontal (medial-to-lateral) parts of the axonal plexus in the ventral nucleus. **C** The terminal, translaminar distribution of horizontally oriented, ventral nucleus axons at lower power. Rapid Golgi method, 12-day-old cat; planachromat, N.A. 0.65, ×500. **D** Probable ascending axonal endings (*1, 2*) in the ventral nucleus and of unknown origin (see text for further discussion). Rapid Golgi method, 25-day-old cat; planachromat, N.A. 1.32, ×1250

terminals are seen at each age studied and occur in mature animals suggests that they are an additional class and not a transitional, ontogenetic stage of other axonal types. These serpentine endings differ from immature axons arising from other sources (Fig. 9: *1, upper left*; Fig. 12B: *4*).

3.5 Cortical Connections of the Ventral Division

Small injections of horseradish peroxidase into the primary auditory cortex (cf. Fig. 5B) produce a narrow band of retrogradely labeled neurons oriented dorso-ventrally in the ventral nucleus, as shown in Fig. 13 (section *85*). In their somatic size and disposition, many of these cells closely resemble the neurons characterized morphologically as principal cells. These and larger injections suggest that the cortical projection of neurons from the ventral and ovoid nuclei is confined to primary auditory cortex (Winer et al. 1977) and to the anterior auditory field next to it (Andersen et al. 1980b). The few retrogradely labeled neurons in the dorsal division are scattered (Fig. 13, sections *72, 60*) or form small clusters without a laminar orientation. The sparser dorsal division labeling may result from encroachment on the overlying cortical areas designated as suprasylvian fringe (Woolsey 1961) (Fig. 5A, C) or in the dorsal zone (Middlebrooks and Zook 1983).

Some recent findings suggest that this picture of ventral division-primary auditory cortex connections may require revision. Thus massive injections of horseradish peroxidase in auditory cortex consistently label more than 90% of the neurons in the ventral nucleus, including many of the smaller cells, although a few small neurons remain unlabeled irrespective of the size of the injection. Because of the size of the injection it cannot be proved that every labeled small cell projects to primary auditory cortex. Nevertheless, the vast majority project to cerebral cortex, suggesting that at least three connectionally distinct classes of neurons may occur in the ventral nucleus, namely, large principal cells with bushy dendritic arbors, but without axonal collaterals; smaller cells with stellate dendritic arbors, extensive local axonal branches, and some axonal collaterals which must exit the nucleus of origin; and a second, much less numerous class of small cells whose axon may be exclusively confined to the ventral nucleus or which may project only subcortically (Winer 1984b). The number of retrogradely labeled neurons is inconsistent with the proportions of Golgi type I (about 60%) and Golgi type II (about 40%) cells routinely encountered in Golgi preparations from the ventral nucleus (Fig. 6) (Morest 1971). The results of transport experiments suggest that each ventral division lamina is represented in the primary auditory cortex and anterior auditory field. These experiments invariably produce small numbers of labeled neurons scattered in the medial division.

3.6 Neuronal Architecture of the Dorsal Division

The dorsal division is the largest of the three parts of the medial geniculate body. It consists of at least five discrete nuclei (and ten parts including the caudal and rostral dorsal nuclei; Table 1) and a large variety of diverse cell

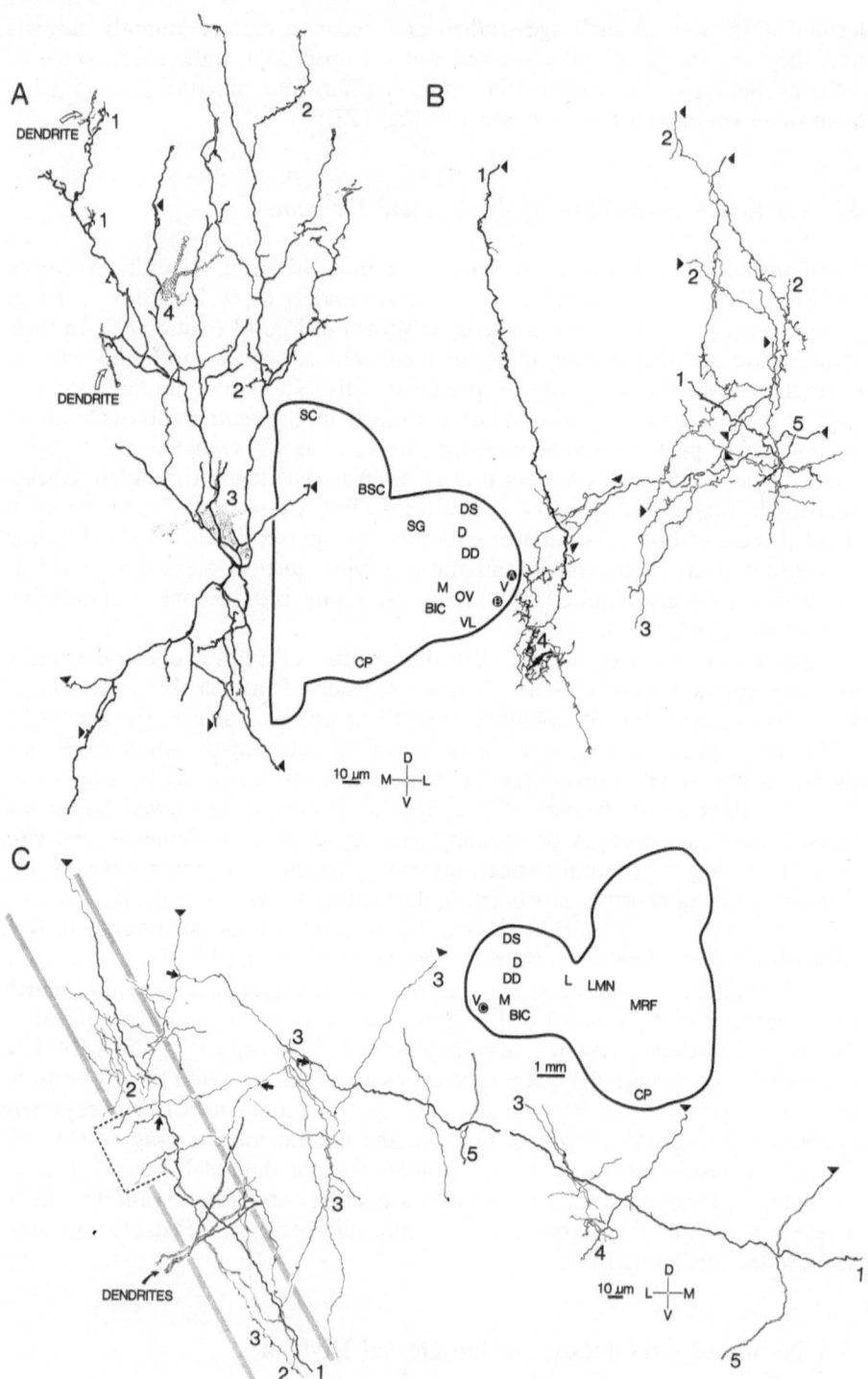

types. It is without the conspicuous laminar structure of the ventral division (Winer and Morest 1978). Dorsal division nuclei have multiple, often overlapping cortical targets (Winer et al. 1977). The dorsal division occupies the entire caudal extremity of the medial geniculate body (Figs. 2A, 3A, 44C), and neurons in this region are rarely labeled after injection of retrograde tracers in primary auditory cortex. Sections rostral to the posterior one-third of the medial geniculate body (Fig. 2B) show the expansion of the ventral and medial divisions and a slight relative reduction in size of the dorsal division. At this level, all the nuclei of the dorsal division are present, and in more rostral sections some nuclei (e.g., Fig. 2C, D: the anterior pole of the ventral division) are reduced in size as the anterior superficial dorsal, anterior dorsal, and anterior deep dorsal nuclei appear.

Each dorsal division nucleus has a distinctive cytoarchitecture and neuronal structure. In Nissl material (Figs. 3, 4) two main areas, which also differ in their structure and connections, are seen. The dorsal nuclei proper consist of three nuclei, each with a posterior and anterior subdivision. The superficial dorsal (Fig. 4D), dorsal (Fig. 4E), and deep dorsal (Fig. 4F) nuclei contain primarily medium-sized and small cells (Fig. 44C), which are somewhat smaller, less densely packed, and distinct from the adjoining ventral division (Fig. 4E, *lower right*). Just behind the caudal extremity of the ventrobasal thalamic nuclei, the anterior dorsal nuclei are present (Fig. 2D). The anterior dorsal nuclei are distinguished by their more stellate dendritic branching pattern, by the well-developed fiber plexus, and by their relative reduction in size compared with the caudal dorsal nuclei.

The second main part of the dorsal division has two subdivisions: the suprageniculate (Fig. 4G) and posterior limitans (Fig. 4H) nuclei. Suprageniculate neurons are the largest cells in the dorsal division (Morest 1964). Neurons in the posterior limitans nucleus form a slender strip on the medial margin of the medial geniculate body, abutting the brachium of the superior colliculus

◁ **Fig. 12A–C.** The terminal fields of extrinsic axons in the ventral nucleus. **A** A large, ascending axon whose collaterals occupy approximately the width of a single laminae and whose terminals pass over bushy principal (*3*) and flask-shaped Golgi type II (*4*) neurons. This axon has collaterals with heterotypic spines (compare *1* and *2*) and other, unusual terminal appendages (*three-sided arrowheads*). (From Morest and Winer to be published) Rapid Golgi method, 41-day-old cat; planachromat, N.A. 1.32, × 1000. **B** Varieties of thinner, extrinsic axons. Some are probably ascending (*1, 4*) and others descending (*2, 3, 5*) by virtue of their trajectory. Most are less than 1-μm-thick and have fine, knobby appendages (*2*) or, more rarely, larger terminal specializations (*4*). Many finer collaterals (*three-sided arrowheads*) leave the plane of this 100-μm-thick section. Rapid Golgi method, 41-day-old cat; planachromat, N.A. 1.32, × 1250. **C** Some features of the preterminal (*right side*) and terminal (*left side*) plexus of various ventral nucleus axons. The thick, ascending axons (*1*) approach their targets and form collaterals (*five-sided arrowheads*) to, and perhaps among, adjacent laminae, while some branches leave the section (*three-sided arrowheads*). The terminals of some knobby axons cross the long laminar axis at right angles and form overlapping and parallel terminal fields (*dashed lines, lower left*) within a lamina (*stippled lines*). Other, probably descending, axons (*3, 5*) are present (see **B**). These axons cross the brachial axons, and their plexus is parallel to the translaminar dendrites (*lower left*). The preterminal part of these axons is as thick as that of ascending fibers, and it has knobby expansions. *2*, fine afferents; *4*, thin axon with serpentine processes. (From Morest and Winer to be published) Rapid Golgi method, 41-day-old cat; planachromat, N.A. 1.32, × 1250

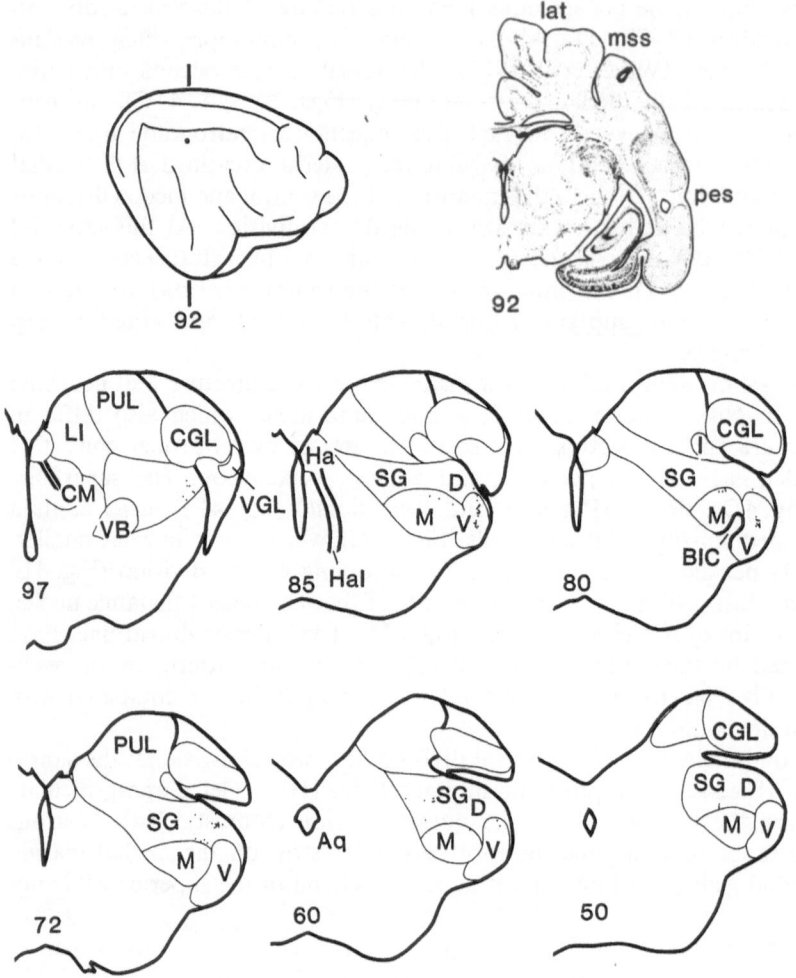

Fig. 13. The distribution of labeled cells in the medial geniculate body following an injection of 0.1 μl of horseradish peroxidase in primary auditory cortex (AI) of an adult cat. The main result is the band of labeled neurons in the ventral nucleus, and a few scattered, labeled cells in the medial division. (Redrawn from Winer et al. [1977])

and the pretectum. The ventrolateral nucleus contains primarily stellate cells but lies within the ventral nucleus (Fig. 1 A).

In the dorsal nuclei four kinds of cells have been described (Winer and Morest 1978, 1983a, 1984). Most numerous are the two large varieties of principal cell whose main targets are the non-primary subdivisions of auditory cortex (Winer et al. 1977). These neurons — the stellate (Fig. 14: *9, 24, 31, 62*) and the bushy (Fig. 14: *4, 5, 13, 49*) cells — have different dendritic configurations which sharply distinguish them from each other and from the cells in the ventral nucleus. A smaller cell with a stellate dendritic domain and a locally arborizing axon is also found throughout the dorsal nuclei (Fig. 14: *8, 12, 37, 50*). The rarest neuron is a medium-sized cell whose axon appears to be intrinsic (Fig. 21). The large stellate, bushy, and small stellate cells occur in about equal numbers.

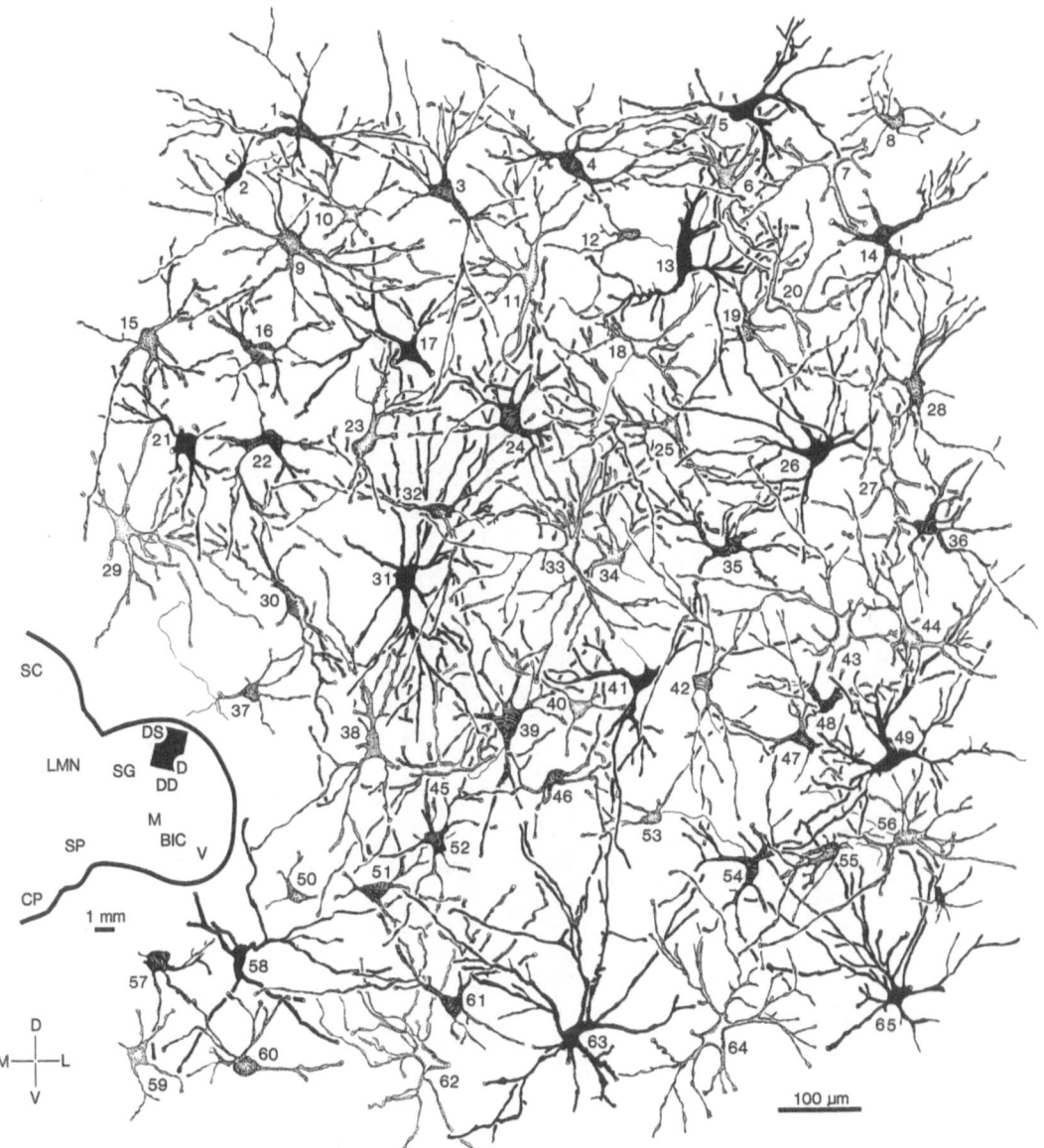

Fig. 14. The varieties of neurons in Golgi preparations from the dorsal nuclei. Prominent are cells with stellate (e.g., *9, 62, 63, 65*) or bushy (e.g., *4, 28, 31, 44*) dendritic configurations. Other neurons include small cells with stellate-shaped dendritic fields (e.g., *8, 12, 37, 50*). There is no obvious tendency for the cells to form laminae, although they do cluster in small groups (see Fig. 4). These cells were drawn from two serial, 100-μm-thick sections. Golgi-Cox method, adult cat; semi-apochromat, N.A. 1.05, × 375

Large stellate cell dendrites occupy a spherical domain in any plane of section (Fig. 15). Four-to-six primary dendrites project from the round or slightly oblate soma. They branch obliquely no more than three times and taper to form delicate terminal segments, some having growth cones in immature cells (Fig. 15: *arrowheads*). Their radiating dendrites cross, but do not coil about, each other as do those of ventral nucleus cells (Fig. 10). In the superficial dorsal

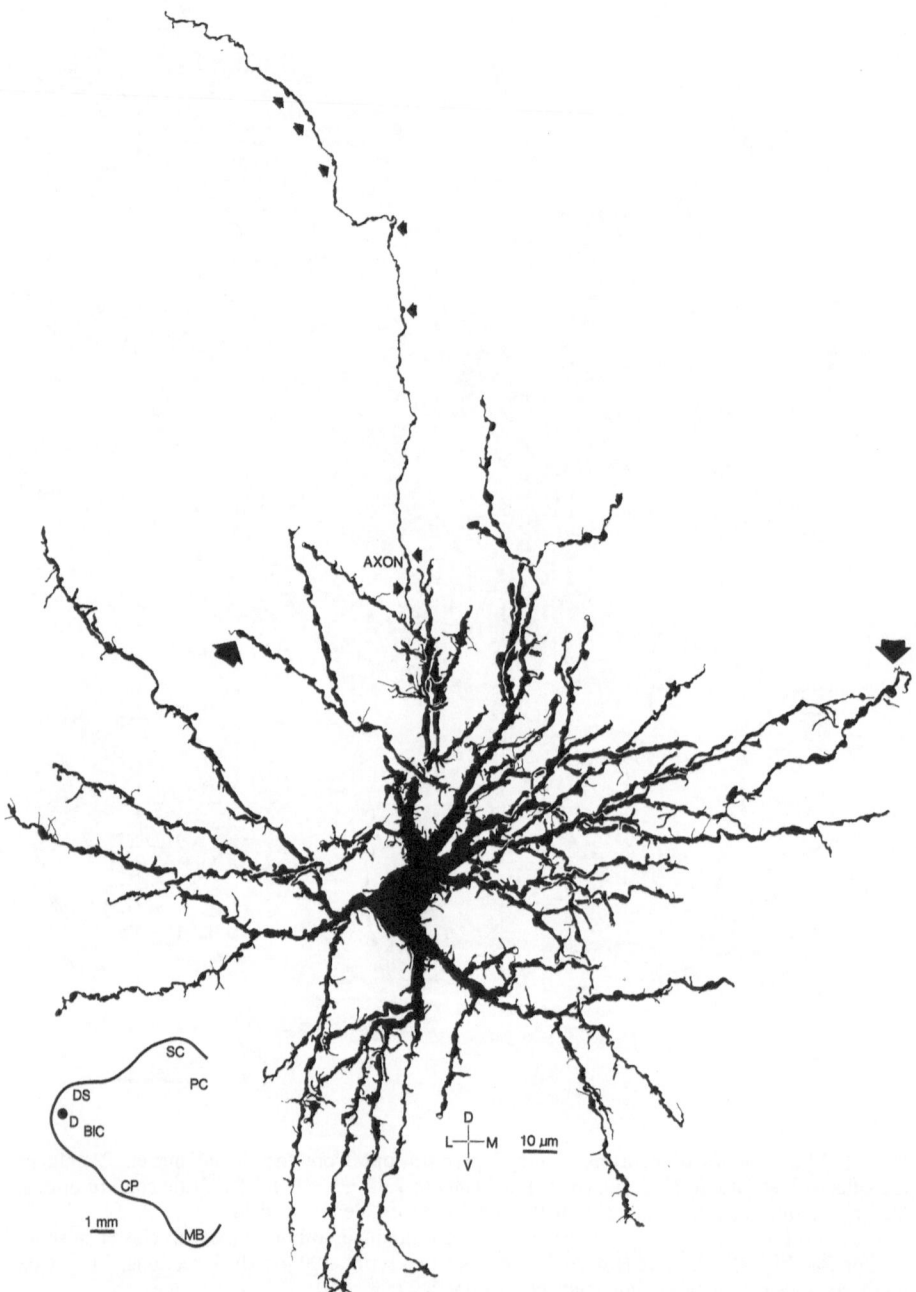

Fig. 15. A stellate principal neuron from the dorsal nucleus. This cell has a radiate, spherical dendritic field about 250 μm in diameter, and the dendrites have stringy, immature spines which will become sparser, smaller, thicker, and pedunculated with age. The axon of such immature cells is usually without collaterals, although numerous swellings (*small arrowheads*) are present. Swellings and stringy appendages like growth cones also occur on dendrites (*large arrowheads*). Rapid Golgi method, 7-day-old cat; planachromat, N.A. 1.32, ×1250

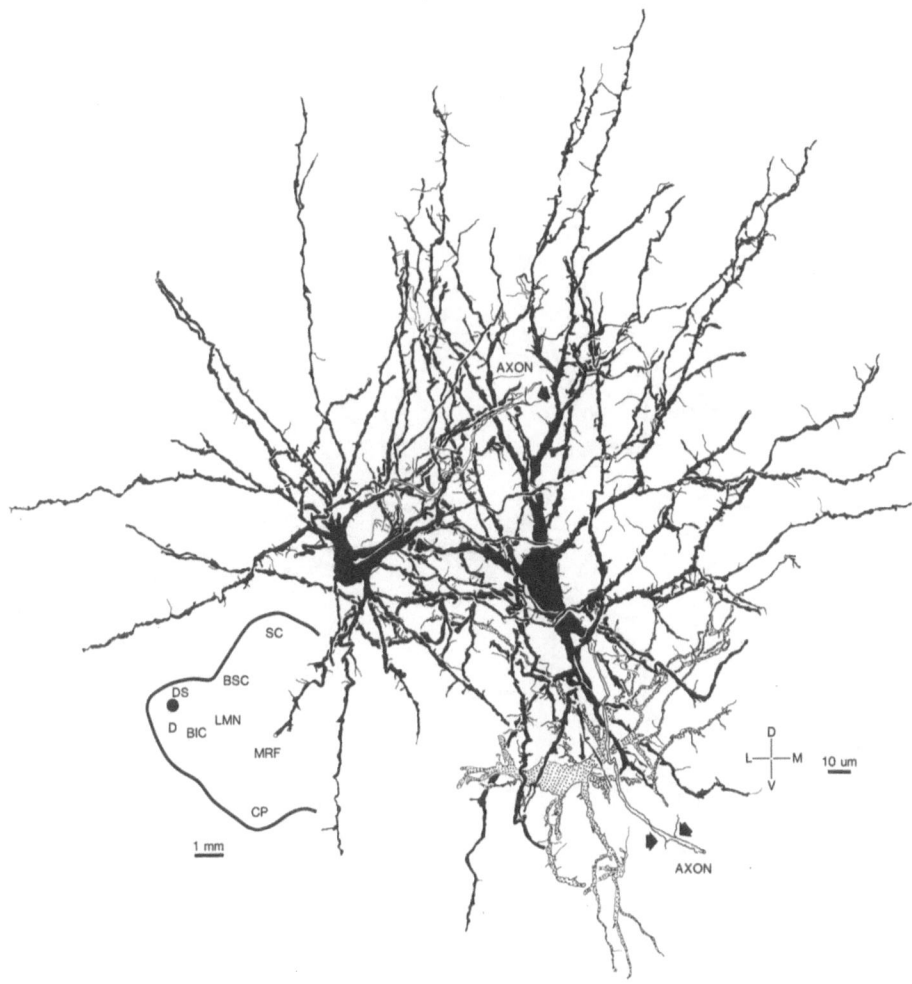

Fig. 16. The interrelation of stellate cell dendrites (*solid black*) in the dorsal nucleus with the dendrites of an adjacent, smaller stellate neuron (*stippled*). These immature principal cells each have a thick axon with a very sparse collateral system (*arrowheads*; see Fig. 15, and Winer and Morest 1983a). Both principal cell and small stellate cell dendrites form densely intermingled nests. Rapid Golgi method, 6-day-old cat; planachromat, N.A. 1.32, ×1250

nucleus, especially near the marginal zone, the dendrites of principal neurons have the appearance of a *noyau fermé* as they conform to the fibrous capsule of the nucleus (Mannen 1960). Except in young animals (Figs. 15, 16), the soma is devoid of appendages, although it may be irregular. The slender dendritic filopodia retract in length but remain relatively constant in number with maturity (Winer and Morest 1983a). Among the loosely packed neurons of the superficial dorsal nucleus (Fig. 2B, C), stellate cells form small clusters. In the dorsal nucleus the stellate neurons are more concentrated, and their density is greatest in the deep dorsal nucleus (Fig. 2B). Here they seem also more intensely stained (Figs. 3, 4A), and the axonal plexus is correspondingly imbricated. Certain patterns in the organization of axons in the dorsal nuclei are correlated with the disposition of cell types (see Sect. 3.7).

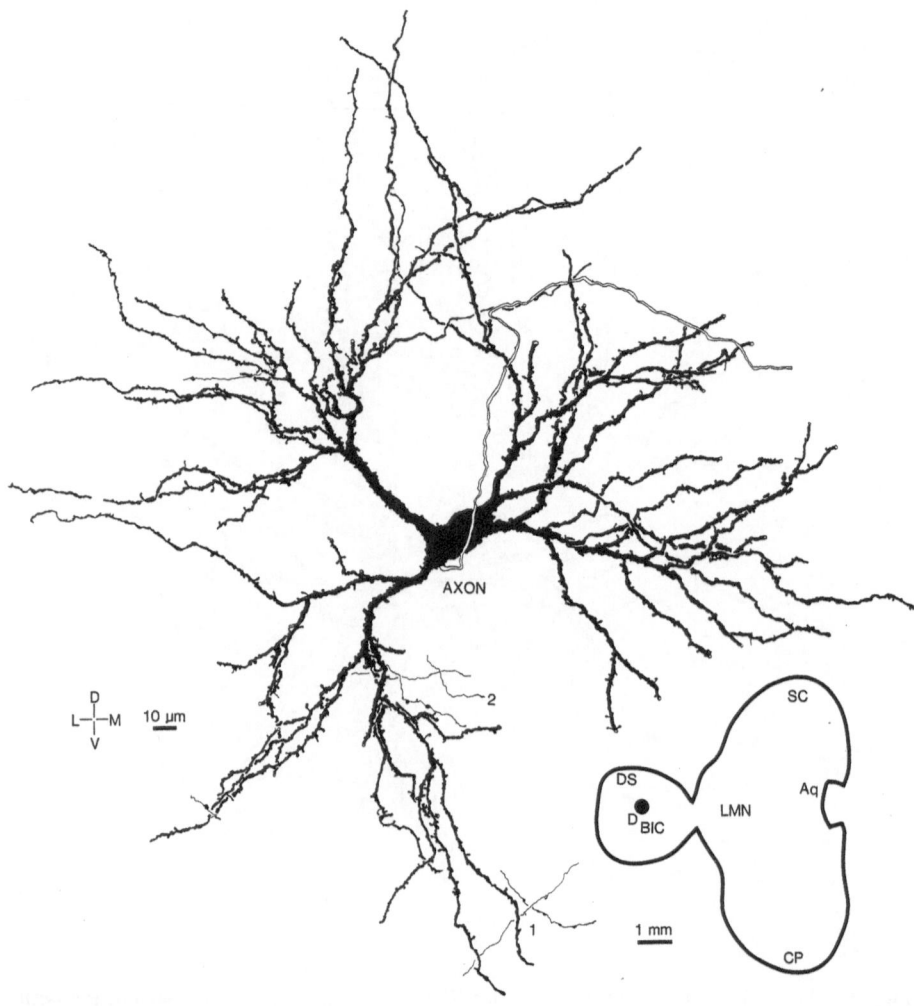

Fig. 17. A bushy principal neuron from the dorsal nucleus. Although it occupies approximately the same volume as the stellate cell (Fig. 16), its dendrites form conical arbors. These cells often have a few and sometimes many somatic spines or spicules. The dendrites have a variety of pedunculated appendages. The thick axon, which is without collaterals, projects toward the acoustic radiations. Some extrinsic axons (*1, 2*) have terminal fields near the central shafts of dendrites. Rapid Golgi method, 23-day-old cat; planachromat, N,A. 1.32, × 1250

Neurons with bushy dendrites comprise the second type of principal neuron in the dorsal nuclei (Fig. 17). They are as numerous as the large stellate cells and are found throughout the dorsal nuclei. They differ from stellate neurons and from every other principal neuron in the dorsal division (except for posterior limitans nucleus neurons – see Sect. 3.8), in that their dendrites are polarized. Though rarely as strongly tufted as ventral division cells (Fig. 6), their three or four large dendrites form conspicuous, conical arbors. In the superficial dorsal nucleus (Fig. 1 A) these bushy dendrites may form poorly differentiated or incomplete layers, but they never form the continuous, laminar pattern evident in the ventral nucleus. In the dorsal, deep dorsal, and anterior dorsal nuclei, no regular stratification of bushy cell dendrites occurs (Fig. 2 B, C).

Each bushy cell's primary dendrite emerges from the flattened (Fig. 17) or triangular (Fig. 18) soma and, about 40–80 μm away, divides into several branches. The central dendrites of these tufts form oblique angles to the main shaft, while the most lateral branches are oriented acutely. Although the shape of the neuron is roughly ovoid in two planes, it is irregular in the third plane: it is not uniformly spherical in each plane, as are stellate cells. After branching two (and occasionally three) times, the dendrites taper, and most end within even relatively thin (100 μm) sections. When they pass from the plane of section, usually an entire tuft is transected (Fig. 18, *lower right*).

The dendritic appendages of bushy neurons resemble those of stellate cells but are usually shorter. The diversity of these appendages (and those of other cell types), when examined at very high power (×2500), is more striking and complex than in previous Golgi studies (Morest 1964). Typical appendages include short or long stumpy spines; flattened excrescences; short or long pedunculated appendages; short or long filiform spines; appendages with forked branches; appendages which make recurrent contact with their parent branch, and other, intermediate, types. These appendages are neither artifactual nor a consequence of maturation; this is confirmed in adult material and in electron micrographs (Morest 1971, 1975a; Winer unpublished results). In addition, the dendritic surface of this and other cell types is not smooth or regular but has numerous indentations and crenations. Bushy cell axons appear similar in structure and trajectory to those of stellate neurons.

The two kinds of interneurons in the dorsal nuclei comprise a third or more of the cells. Their many axon collaterals and the plexus of extrinsic axons impart a characteristic texture on the axonal plexus of the dorsal nuclei. By far the most numerous are the *small stellate Golgi type II cells* (Fig. 19). The 10 × 15–20 μm, oval perikaryon has four-to-six dendritic branches which project in an irregularly spherical pattern. Their radiate dendritic fields are a fraction of the volume of those of principal cells. Most dendrites end within the plane of section, and a few are extraordinarily long (Fig. 20 A). The thin trunks usually branch fewer than three times. Their appendages are sparse, variable in shape, thin, and often pedunculated, and the surface of the dendrite has many irregular swellings. Near their ends the dendrites may branch again, and their distal segments often taper to a diameter of 1 μm or less.

The most striking feature of small stellate cells is their complex system of axon collaterals (Figs. 19, 20 A). Emerging from the soma or a dendritic trunk, the axon branches repeatedly (up to 80 times or more) and fills an area about equal to the dendritic field of the cell. Collaterals arise acutely from the main branch and follow a tortuous, corkscrewlike path through the neuropil (Fig. 19: 1), and have many complex terminal patterns (Fig. 19: 2–4). Terminal branches are often extremely fine — less than 0.5 μm thick — and among the thinnest axons in the medial geniculate body. The collateral network is more extensive and has more branches than those of flask-shaped Golgi type II cells in the ventral division (Morest 1971, 1975a). The small stellate cells in the dorsal nuclei are common near stellate (Fig. 16, *stippled outline*) or bushy (Figs. 18, 20 A, *hatched outlines*) principal neurons. Their axons here form dense, peridendritic arbors. The Golgi type II axons often have recurrent collaterals

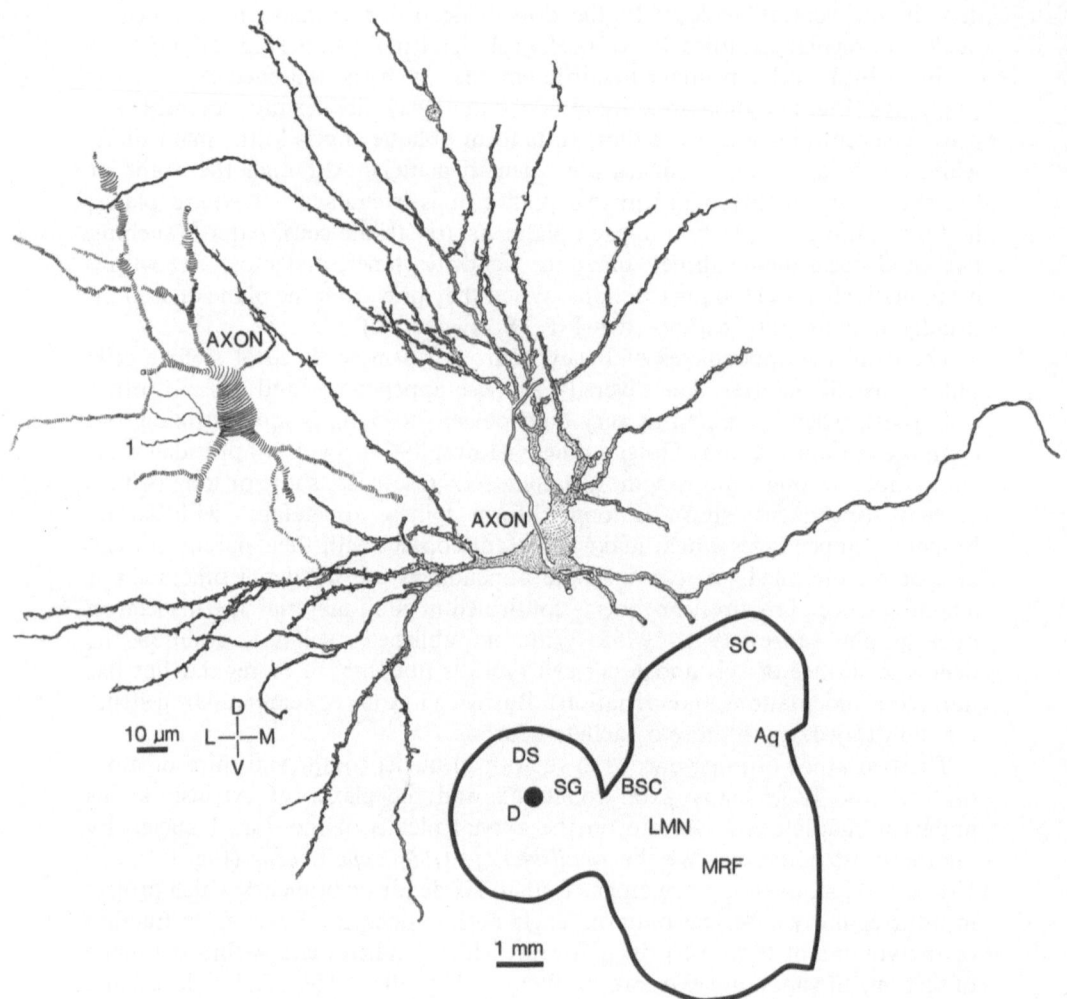

Fig. 18. The interrelations between a small stellate neuron (*hatched*) and a bushy principal neuron (*stippled*) from the dorsal nucleus. These cell types are so frequently apposed that this arrangement could be a basic feature of dorsal division organization. A few thin, beaded axons (*1*) pass near the stellate cell's dendrites, and the axon of this cell appears to contact the central dendrites of the principal cell before leaving the plane of section. Often the dendrites of these cells wind around each other. Rapid Golgi method, 23-day-old cat; planachromat, N.A. 1.32, × 1250

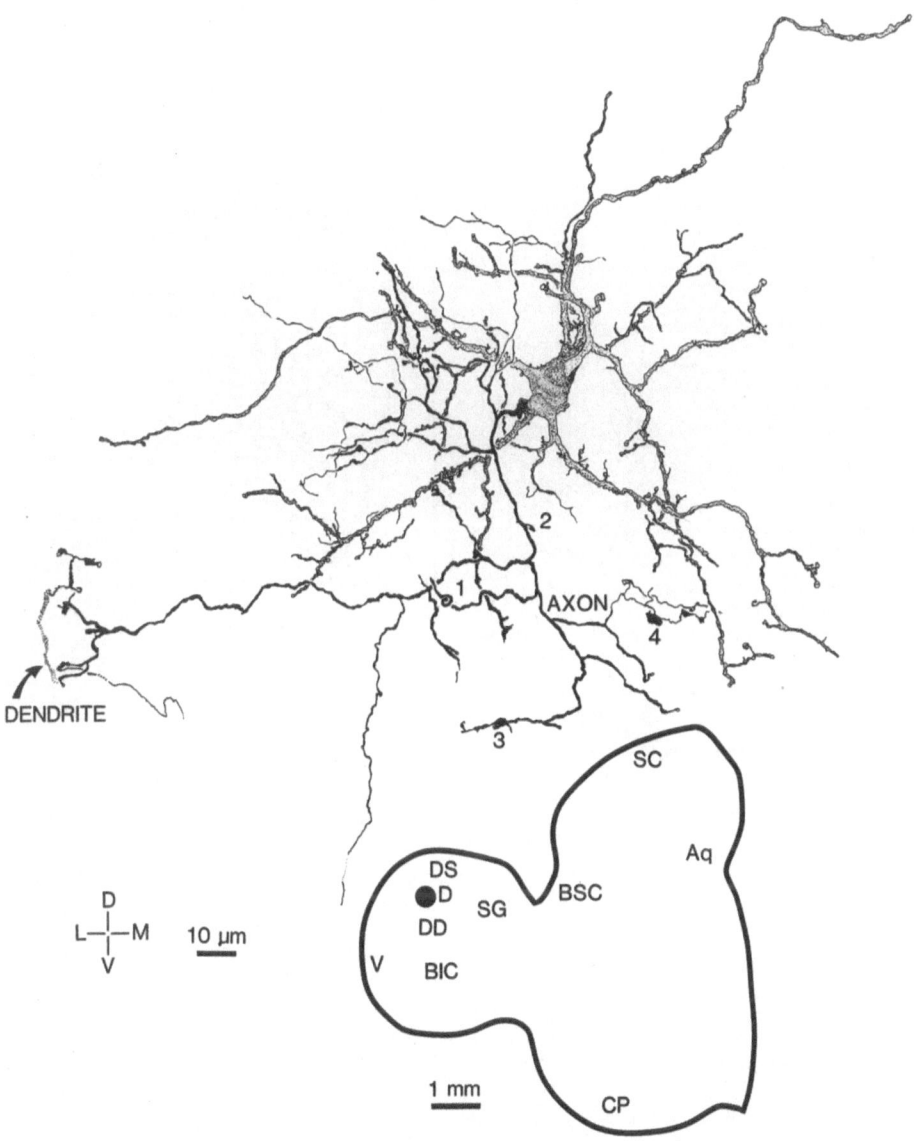

Fig. 19. A small stellate Golgi type II cell from the border of the superficial dorsal and dorsal nuclei. The slender, sparsely spinous dendrites fill a limited, irregular area. This cell is remarkable for the many extremely fine axon collaterals and their complexity. The collateral plexus dominates the texture of the neuropil in the dorsal nuclei. The axon projects locally and terminates in corkscrewlike arrangements (*1*), short spicules (*2*), slender twiglike buds (*3*), or delicate (less than 0.5-μm thick) collaterals (*4*). Most branches end in the section, but some project outside it. Other, even more elaborate endings also occur (*extreme left*), sometimes near dendrites. Some collaterals project recurrently to the parent cell's dendrites to form fine arbors. The cell shown here has fewer dendritic spines than most small stellate interneurons. Rapid Golgi method, 23-day-old cat; semi- and planapochromat, N.A. 1.25, 1.32, × 1250

37

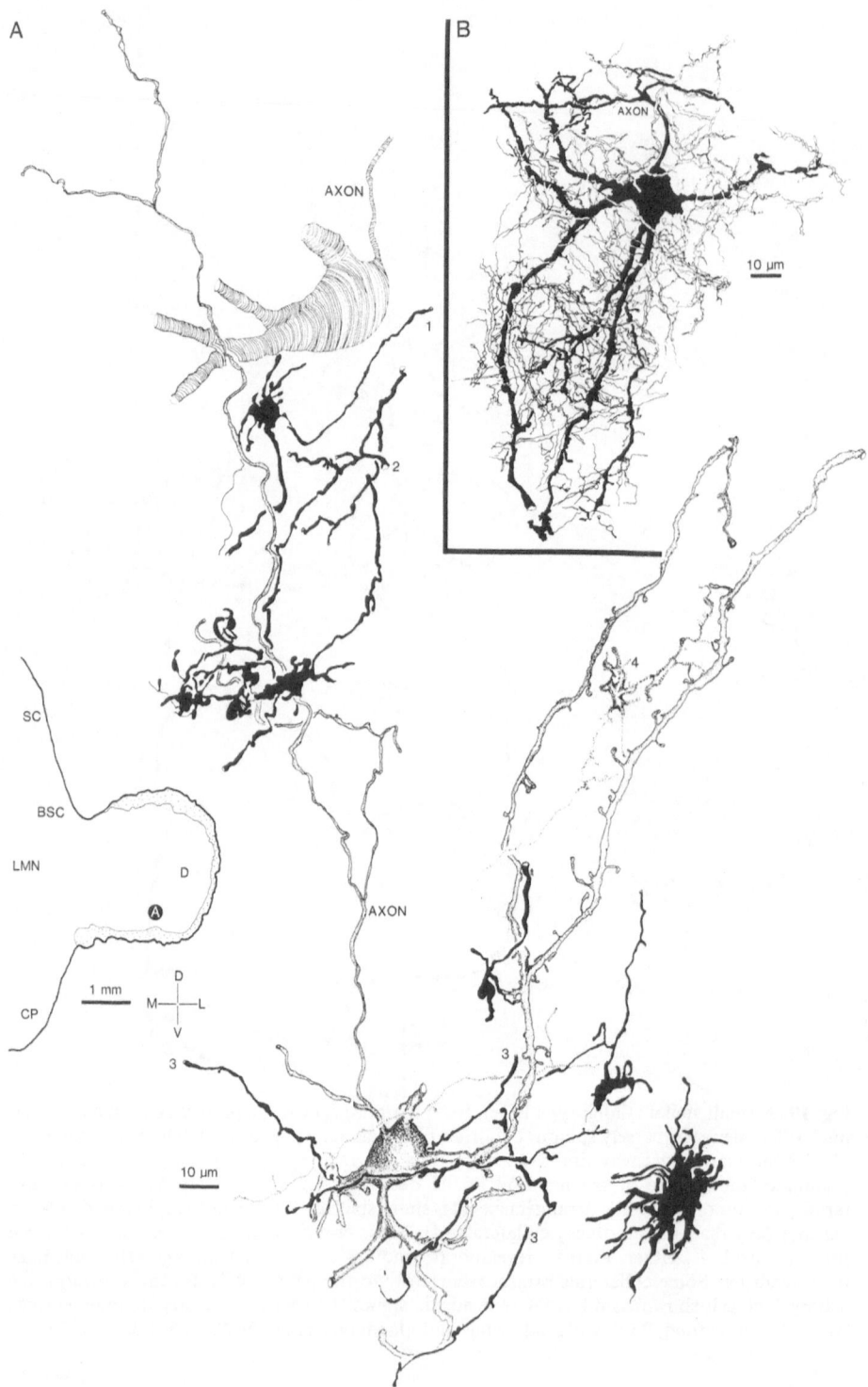

projecting toward the central and most heavily spinous parts of the parent cell's dendrites (Fig. 19, *lower left*). Small stellate neurons occur throughout the dorsal nuclei, including the anterior dorsal nuclei.

A second, much rarer, Golgi type II cell is also found. This *medium-sized interneuron* (Fig. 21) occurs primarily in the superficial dorsal and dorsal nuclei and less often in the deep dorsal nucleus and anterior dorsal nuclei. It is the least common cell type in the dorsal division and probably comprises only a small fraction of its cells. This distinctive neuron has a small, rectangular or elongated soma with four-to-seven primary dendrites. Some of these are thicker and considerably longer than small stellate Golgi type II dendrites. The larger dendrites are encrusted with elaborate appendages or may be virtually spine free. Many kinds of appendages are found. The most striking are long and delicate and have diverse shapes, including multiple branches and complex, interwound segments (Fig. 21, *upper left*). These neurons are frequently found in close association with principal cell dendrites (Fig. 21: *6*). The axon of the large Golgi type II cell resembles the small stellate cell's axon except that it is thicker and may project farther, since many branches usually pass from the plane of section. The axon typically lacks the elaborate local collaterals and complex recurrent systems of small stellate neurons (Fig. 19).

3.7 Structure of Axons in the Dorsal Nuclei

The dorsal division contains a complex network of extrinsic axons which is summarized in Fig. 25. The dorsal nuclei proper and the other parts of the dorsal division are separately considered.

Among the most numerous axons in the dorsal nuclei are ascending type I fibers which enter the medial geniculate body from the brachium of the inferior colliculus (Fig. 22). Both ascending and descending branches (Fig. 22A) ramify and form collateral nests which are primarily peridendritic (Fig. 22C). Clusters of these slender (about 1-μm-thick) collaterals form a dense plexus throughout the dorsal nuclei, and these axons (Fig. 25A: I; Fig. 25B: I) have many different appendages (Fig. 22B). Their structure and sinuous course resemble tendrils of ivy. Although the main target of type I axons is the spinous middle segments of principal stellate (Fig. 22C) and bushy (Fig. 17: *1*) cell dendrites, they probably have other projections too. These include small stellate Golgi type II dendrites (Fig. 20A: *hatched axon*) where they ramify to form a dense plexus (Fig. 20B). They also cluster near the dendrites of large Golgi type II cells (Fig. 21: *3*). A second common fiber is a thicker variant of the type I axon. This type II

◁ **Fig. 20A, B.** A small stellate Golgi type II cell from the caudal extremity of the dorsal nucleus and showing the varieties of extrinsic axons. **A** The stellate cell axon (*stippled*) projects toward the dendrites of a nearby bushy neuron (*hatched*). Among the axons are type IV fibers (*1*, *4*), type II axons (*3*), and a type VII ending (*2*). An oligodendrocyte is shown (*solid black, lower right*). Rapid Golgi method, 41-day-old cat; semi-apochromat, N.A. 1.25, × 1250. **B** The axonal plexus adjacent to a Golgi type II cell. Some thick axons (*hollow outlines*) originate from nearby Golgi type II cells. The fine axons are from types I, II, III, V, and VIb (see Fig. 25). (Modified from Winer and Morest 1984) Rapid Golgi method, 25-day-old cat; plana-pochromat, N.A. 1.32, × 1250

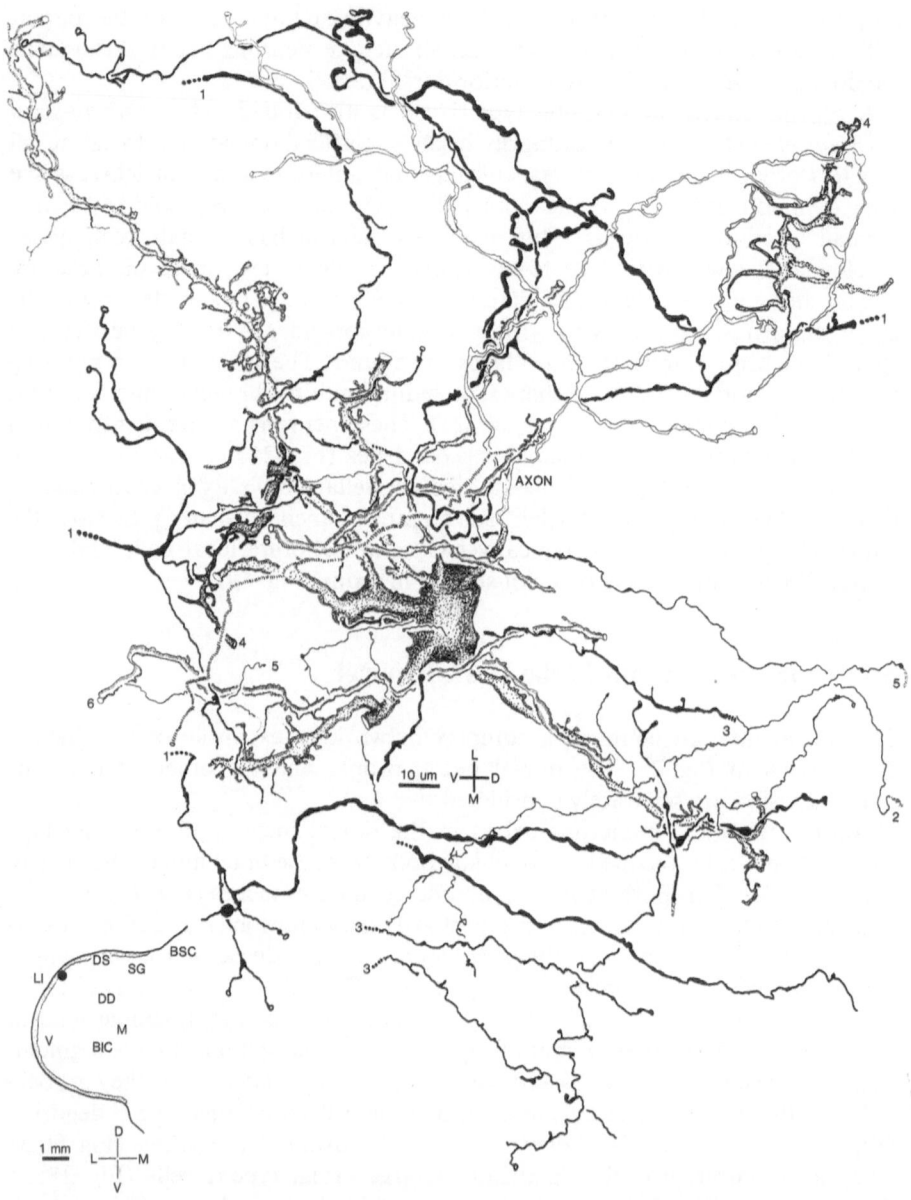

Fig. 21. A large stellate Golgi type II cell from the superficial dorsal nucleus and some nearby axons. The axons of adjacent Golgi type II cells (*1*) are common, and many form axospinous terminals. Type V axons (*2*) are also present, as are type IV axons (*4* and *upper right*). Some beaded type VIb axons arborize in the neuropil (*3*). Perisomatic arbors, possibly from type II axons, are present, as well as thin, beaded, and perhaps descending fibers (*5*). Principal cell dendrites also pass through this area (*6*). (Modified from Winer and Morest 1984) Rapid Golgi method, 41-day-old cat; semi-apochromat, N.A. 1.32, ×1250

fiber (Fig. 25 B: II) has a similar course and form and shares many of the targets of type I axons (e.g., small Golgi type II cells: Fig. 20 B).

Type III axons (Fig. 25 B: III) are extremely slender and enter the dorsal division dorsally and medially. Their sparse collateral systems form delicate nestlike terminal buds near principal cell dendrites (Winer and Morest 1984).

The most distinctive axon in the dorsal nuclei is the type IV (caliciform or grumous) ending (Fig. 25 A, B: IV). This coarse axon enters the dorsal nuclei ventromedially, from the brachium of the inferior colliculus. It has several subsidiary branches as thick as, and sometimes thicker than, the main trunk (Fig. 23 A). These collaterals branch extensively, often in clusters, and form unique, peridendritic grumous specializations throughout the dorsal nuclei. The terminal territories of type IV axons are the largest and most complex of any axon in the dorsal division (Fig. 23 A). The caliciform ending consists of a large central portion, the body, often found near principal cell dendrites (Fig. 23 B). The body contains multiple lacunae, and each branch has a collateral plexus which can extend for some distance. The grumous segment of the caliciform ending may have one central part and a modest (Fig. 23 C: *4, 5*) or enlarged collateral system (Fig. 23 C: *3*), or be enormously expanded, with many elaborate specializations (Fig. 23 C: *2*). That these endings appear in mature animals and lack the delicate qualities characteristic of immature axons suggests that they are a separate, and not a transitional, ontogenetic form. Further, they are found with other classes of axons. A major target of type IV axons is the dendrites of principal stellate and bushy neurons (Fig. 20 A: *1*; Winer and Morest 1984). These axons also end near the dendrites of small stellate Golgi type II cells (Fig. 20 A: *4*) and large Golgi type II cells (Fig. 21: *4*).

Type V axons (Fig. 25 A: V; 25 B: V) enter the medial geniculate body from its dorsal, parabrachial border and pass obliquely through the dorsal nuclei, where they form small fascicles. These axons have slender, preterminal parts which branch sparsely to form a diffuse plexus in the dorsal nuclei. These axons may be part of a descending system in the dorsal nuclei.

Type VI fibers are thin (type VIa: Fig. 25 B) or thick (type VIb: Fig. 25 B) axons with beadlike swellings and sparse collaterals along their length. They are from 1–3 μm thick and enter the medial geniculate body from its anterior, dorsal, and (to a lesser extent) medial aspect. They describe a gentle dorsomedial-to-ventrolateral arc. Their pathway through the superficial dorsal nucleus enhances the quasi-laminar appearance of this nucleus, where the fibers often form strata. These axons are numerous in the dorsal nuclei and end near the dendrites of principal bushy cells (Fig. 17, *above 1, 2*; Fig. 18: *1*) and large Golgi type II cells (Fig. 21: *5*). Their size, ubiquity, and course suggest they are of descending origin (Winer and Morest 1984).

Type VII axons (Fig. 25 B: VII) are thick and have sparse, coarse collaterals. Though not common, their anterodorsal trajectory suggests they may be an ancillary, descending system to the dorsal nuclei. Type VIII fibers (not illustrated) enter the medial geniculate body rostrally, project toward principal cell dendrites, and may also be descending (Winer and Morest 1984). Types I, II, and IV enter the medial geniculate body ventromedially and may be of midbrain origin. Types III, V, and VI enter from the parabrachial regions along the medial aspect or from the brachium of the superior colliculus. The eight varieties (including nine types) of extrinsic axons described above, along with the geniculofu-

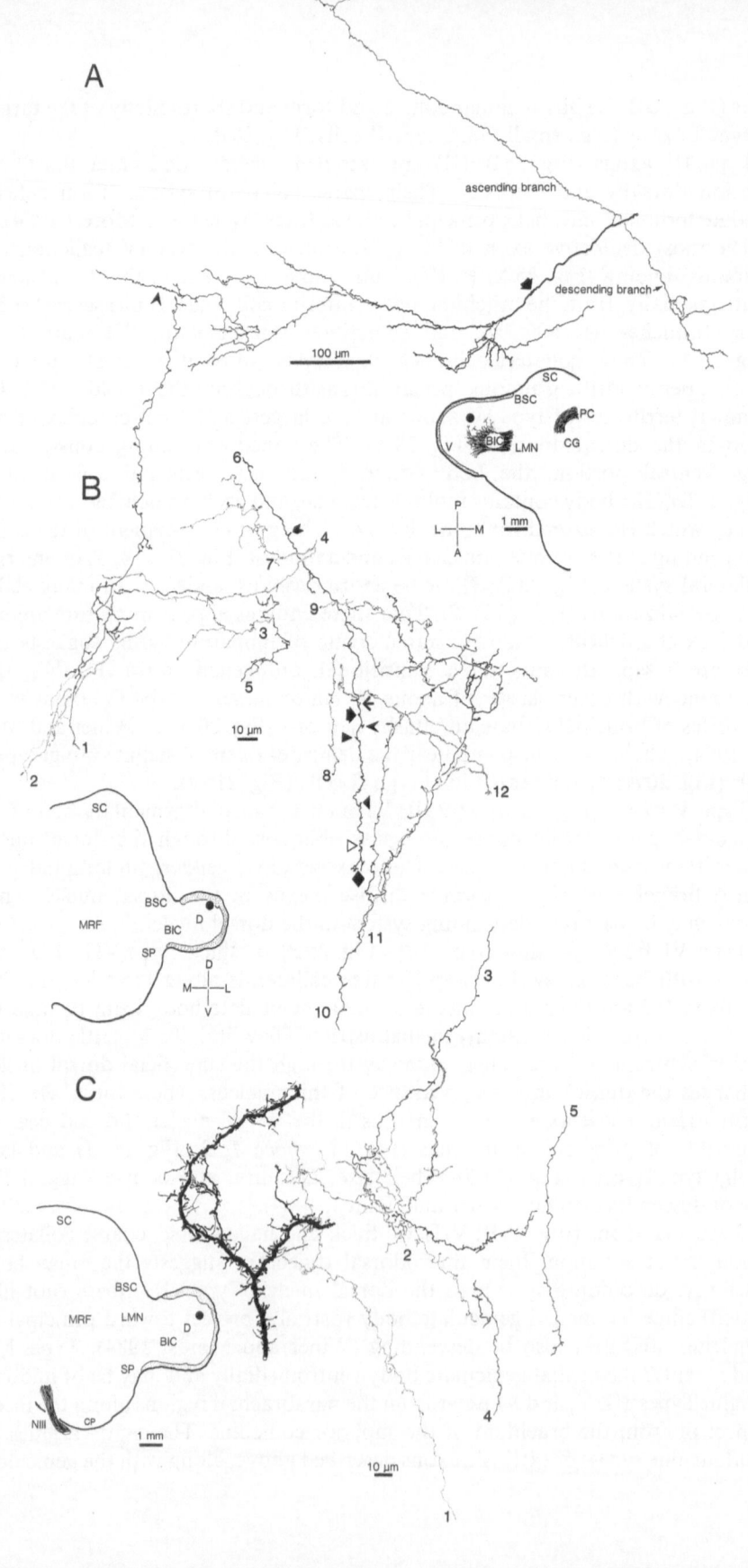

A

ascending branch

descending branch

100 μm

SC
BSC
PC
V BIC LMN
CG

P
L——M
A

1 mm

B

6

4
7
9

3
5

10 μm

1
2

SC

BSC   D
MRF   BIC
SP

D
M——L
V

8

12

11

3

10

C

SC

BSC
MRF LMN D
BIC
SP

NIII   CP

1 mm

5

2

4

1

10 μm

gal axons of the two types of principal neurons and the two kinds of Golgi type II axons, constitute a total of at least thirteen types of axons in the dorsal nuclei. Collectively, they confer a fine, dense texture, without apparent preferential orientation (at least in the dorsal and deep dorsal nuclei) to the dorsal nuclei. The dominant element in the neuropil, and conspicuous for their fine caliber, is the collaterals of Golgi type II axons (Figs. 24, 45B, D), which are also apparent in electron micrographs (Winer and Morest 1984). This contrasts with the comparatively regular, laminar orientation of the ventral nucleus (Figs. 8 B–D, 45A, C). Here the neuropil is dominated by extrinsic axons and less so by intrinsic fibers. On a regional basis, as shown in rapid Golgi and Woelcke-stained material, the neuropil of the dorsal nuclei is most densely stained, respectively, in the anterior deep dorsal, then in the anterior dorsal and anterior dorsal superficial nuclei, followed by the caudal dorsal nuclei in the same order.

3.8 Neuronal and Axonal Architecture of the Suprageniculate Nucleus and the Posterior Limitans Nucleus

The suprageniculate nucleus is an oval mass of predominantly large neurons forming the dorsomedial margin of the medial geniculate body (Fig. 1). Consisting mainly of modified stellate neurons with radiate branching patterns (Fig. 2B–D), it begins at about the midpoint of the caudal dorsal nucleus, a few hundred micrometers from the caudal pole (Fig. 2A). Its neurons continue rostrally (Fig. 3) and consist of loosely packed (Fig. 4A), large and small multipolar cells (Fig. 4G). Its dorsal border is formed by the free surface of the medial geniculate body; medially, by the brachium of the superior colliculus, posterior limitans nucleus, and the lateral mesencephalic nucleus; laterally, by the dorsal nuclei; and ventrally, by the medial division.

Except for the magnocellular neurons of the medial division (see Sect. 3.2), the *principal neurons* of the suprageniculate nucleus are the largest cells in the medial geniculate body (Fig. 26). Their oval or rounded, 30×35 μm perikaryon has six-to-nine major dendrites. Each dendrite divides obliquely two or three times, some terminal branches forming trident-shaped arbors. The dendritic field of this cell is irregularly spherical and the branching pattern radiate. The dendrites project in several planes, and the distal segments often cross obliquely.

◁ **Fig. 22 A–C.** Type I axons in the dorsal division. **A** Most type I fibers have ascending and descending branches which form fascicles throughout the dorsal nuclei. The primary and secondary branches both follow a sinuous course, and diverse appendages (*arrowheads*) are present. Rapid Golgi method, 5-day-old cat; planachromat, N.A. 0.65, × 500. **B** High-power views of type I terminals. Ascending (*10*) and descending (*6*) branches occur, and have a variety of appendages. These include thick terminals with stumpy crowns (*6, short triangular arrowhead*) or serpentine, abruptly curving segments (*6, hollow triangular arrowhead*), and many others (varieties of *arrowheads* and *arrowtails*). Rapid Golgi method, 41-day-old cat; planapochromat, N.A. 1.32, × 1250. **C** Type I terminals near a principal stellate cell's dendrite (*solid black*). Multiple contacts are made by slender branches (*1*), and some terminals coil and form peridendritic nests (*2–4*). Note the long, immature appendages on the dendrites. *5*, thin, probably descending axon. (Modified from Winer and Morest 1984) Rapid Golgi method, 41-day-old cat; planapochromat, N.A. 1.32, × 1250

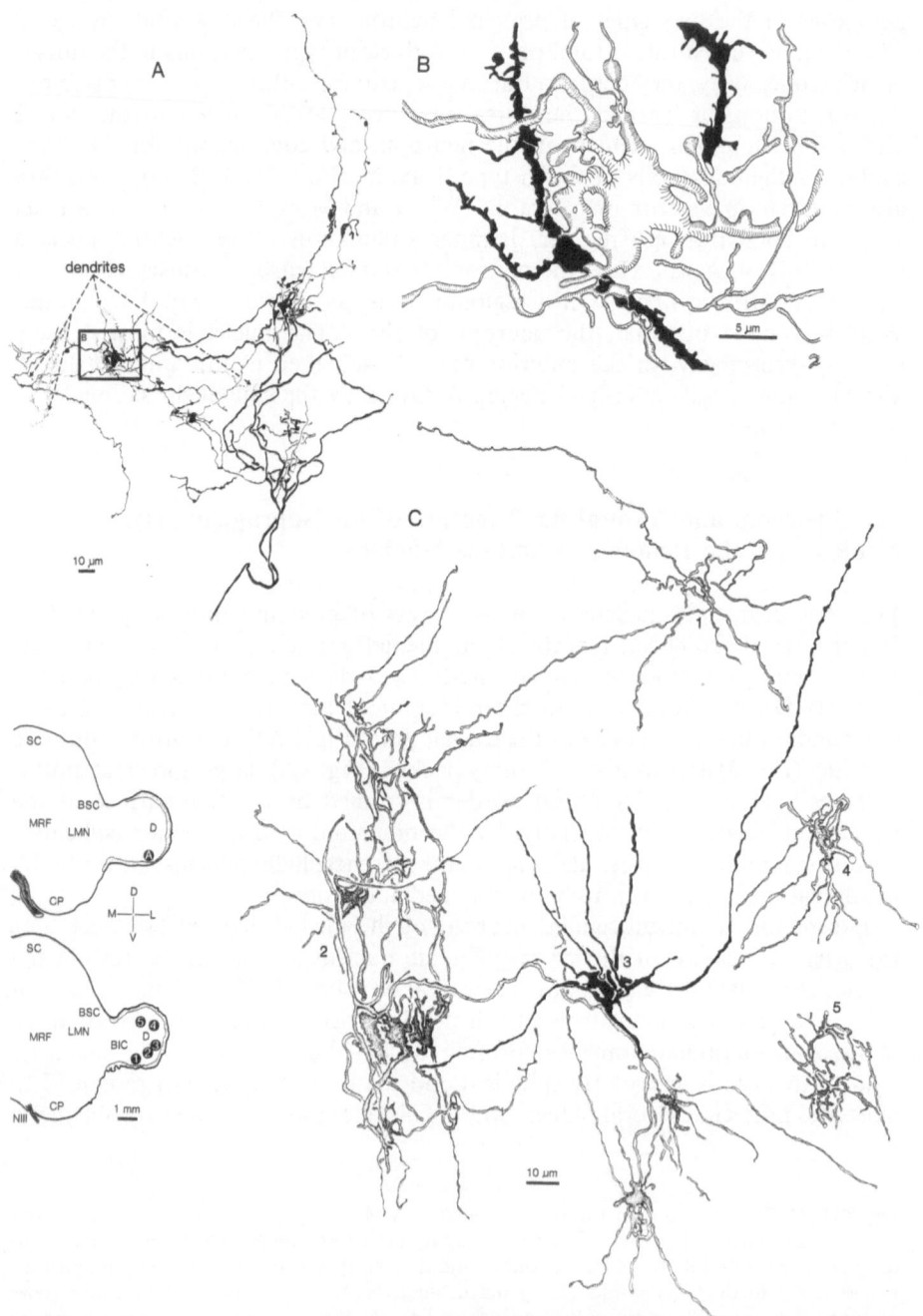

Fig. 23A–C. Type IV grumous terminals. **A** Part of the terminal plexus of a single ending in the caudal pole of the dorsal nucleus. **B** The central bodies of the terminal (*hatched outlines*) are often found near principal cell dendrites (*solid black outlines*). Note the lacunae of the central body. **C** Varieties of type IV axons (*1–5*). Some have multiple central bodies (*2*) or a single central body (*4, 5*). (From Winer and Morest 1984) Rapid Golgi method, 41-day-old cat; planapochromat, N.A. 1.32, ×1250

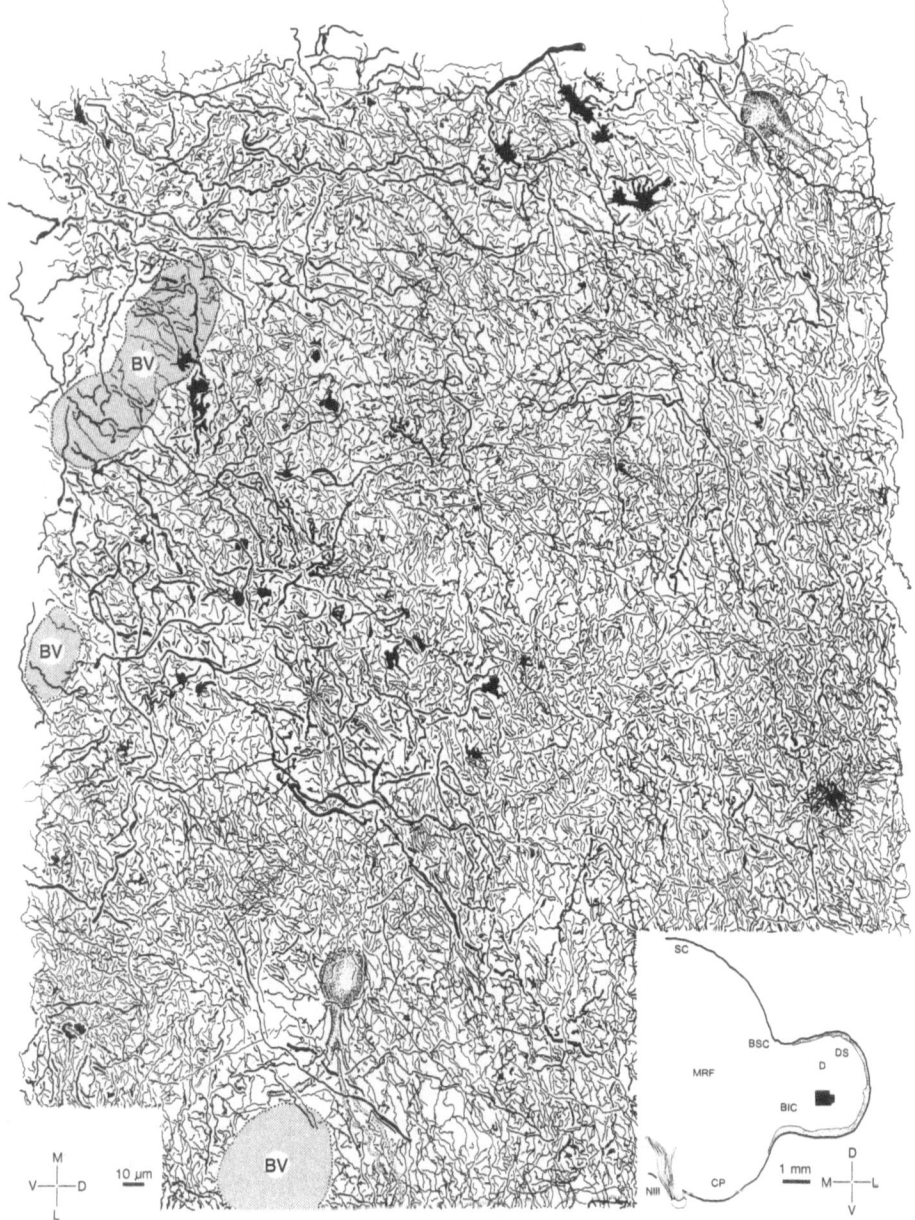

Fig. 24. The axonal plexus of the caudal part of the dorsal nucleus. A variety of axons are present in this 100-μm-thick section, including those from types I, III, IV, and VIb (see Fig. 25). Portions of two bushy cells (*stippled*) occur and small zones of rarefaction in the neuropil (*center*) may correspond to unimpregnated somata. A second source of axons is the fine, intrinsic plexus attributed to Golgi type II collaterals (see Figs. 19–21). The thicker, unbranched endings are probably from principal cells. (From Winer and Morest 1984) Rapid Golgi method, 41-day-old cat; planapochromat, N.A. 1.32, × 1250

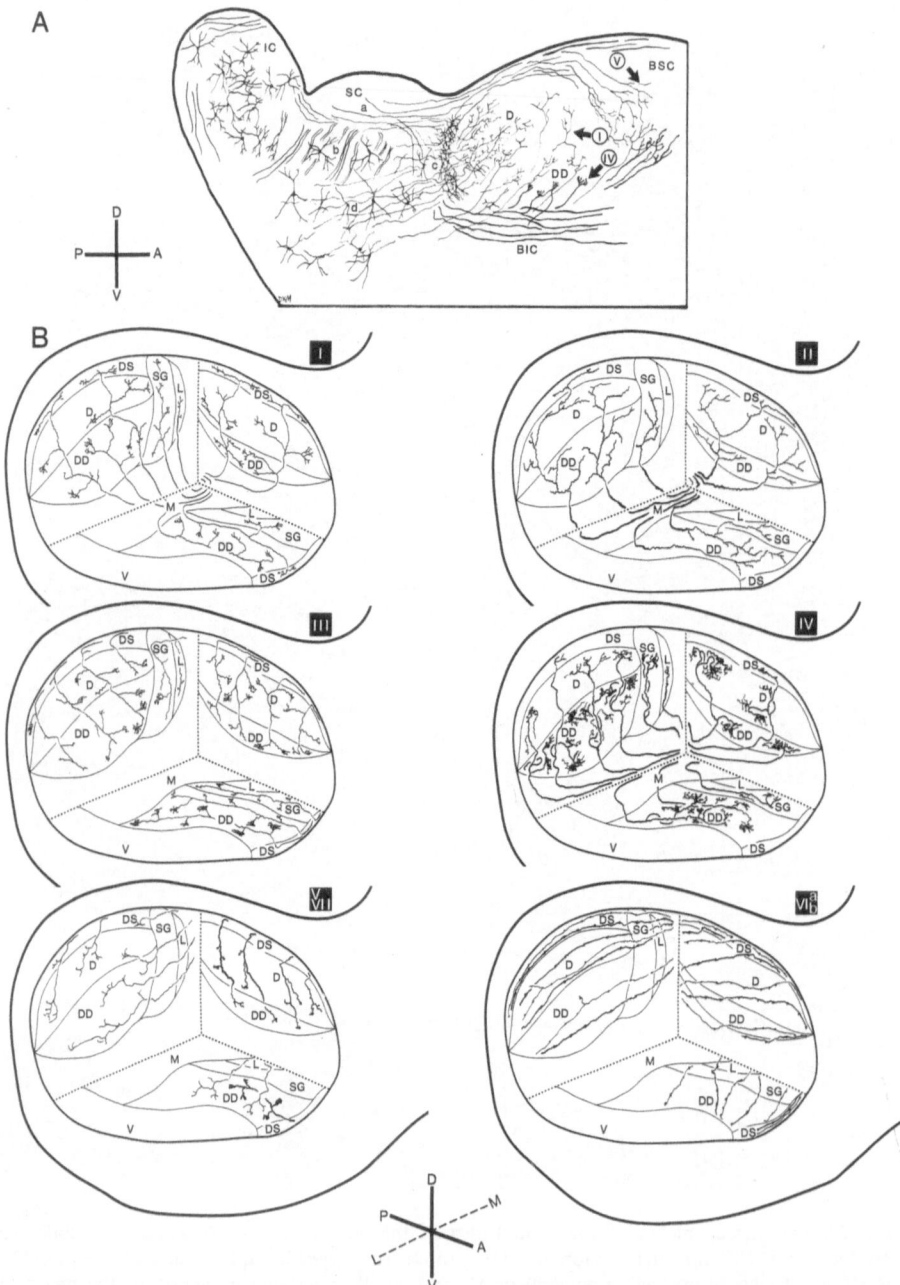

Fig. 25 A, B. A summary of the eight varieties of extrinsic axons in the dorsal division. **A** Parasagittal view of the brain stem redrawn from Morest (1965b) and on which the nomenclature of the present study (and from Winer and Morest 1978, 1984) is superimposed. Axon terminals from types I, IV, and V are present. Golgi-Cox method, 28-day-old cat; Zeiss drawing apparatus. **B** Schematic views of the extrinsic axons in axonometric perspective, reconstructed from rapid Golgi preparations. Note that types I, II, and IV are of brachial origin, and that types III, V, VI, and VII are extrabrachial. Type VIII is not shown here. For types V, VII: left panel, type V; right panel, type VII; ventral panel, both groups. For types VI a, b: left panel, type VI a; right panel, type VI b; ventral panel, both groups. See text for further discussion. (Modified from Winer and Morest 1984)

For a cell this size the dendrites are uncharacteristically thin, less than 4 μm in diameter. They have a variety of appendages; most of these are short but others are like those on principal neurons in the dorsal nuclei. The axon is 2–3 μm thick and comes usually from the soma. It projects anterolaterally, toward the acoustic radiations. This axon has no collateral system and usually acquires a myelin sheath within 20 μm of its origin.

A second, and much smaller, neuron is also found in the suprageniculate nucleus (Fig. 26: *stippled outlines*). This *Golgi type II cell* has either an oblate or flask-shaped soma and three-to-six primary dendrites. These may be as thick as those of principal cells, and they often commingle, as do their axons, among principal cell dendrites. The interneuron dendrites have few appendages, and these are shorter and simpler than those on corresponding Golgi type II dendrites in the dorsal nuclei. The axons of these cells also recapitulate the pattern of projection in the other parts of dorsal nuclei. Most axonal branches pass toward adjacent principal cells, and at least one collateral projects recurrently toward the central part of the dendritic tree. Compared with the dorsal nuclei interneurons, suprageniculate Golgi type II cells have fewer axonal branches and relatively simpler collaterals.

The extrinsic axons in other parts of the dorsal division occur also in the suprageniculate nucleus (Fig. 27), with some exceptions (Winer and Morest 1978, 1984). These include type I "ivy tendril" axons (Fig. 27 A: *3*), type IV calyceal terminals (Fig. 27 B), and thick type VII fibers (Fig. 27 A: *4*). It is difficult to be certain if each axon entering the suprageniculate nucleus terminates there, since many are clearly en route to more lateral targets. The tortuous course and complex collateral systems of many extrinsic axons (Fig. 27 A: *5*) suggest that not all axons entering the suprageniculate nucleus end there.

The texture of the axonal plexus in the suprageniculate nucleus reflects these patterns (Fig. 28). The neuropil is somewhat less uniform than in the dorsal nuclei (Fig. 24) and is largely dominated by axons oriented medio-laterally or traversing the nucleus. The structure of individual axon terminals in the suprageniculate nucleus is often simpler than that in adjacent dorsal division nuclei (e.g., the size of the type IV caliciform endings: Figs. 27 B, 28). The number and complexity of local circuit collaterals may also be reduced, and there are areas of comparative rarefaction in the neuropil not present in the dorsal or deep dorsal nuclei.

The posterior limitans nucleus is a slender lamella wedged between the suprageniculate nucleus and the adjoining brachium of the superior colliculus and lateral mesencephalic nucleus (Fig. 1). It is the smallest nucleus of the dorsal division and is present only between the caudal extremity (Fig. 2 B) and the midpoint of the medial geniculate body. In Nissl-stained sections it is conspicuous, and its somata are oriented in a more or less dorsolateral-to-ventromedial direction (Fig. 4 A, H). The larger somata are preferentially aligned along this axis and are rather uniform in size. The neuronal architecture of the posterior limitans nucleus confirms this lamellar structure (Fig. 29). The elongated, almond-shaped *principal cell* somata has two or three major dendrites which branch usually only once or twice. Most dendrites parallel the long axis of the somata although a few cross it a right angles. These cells resemble certain sparsely branched neurons in the reticular formation (Ramon-Moliner and Nauta 1966). The soma is about 30 × 15–20 μm in size, and the dendrites are

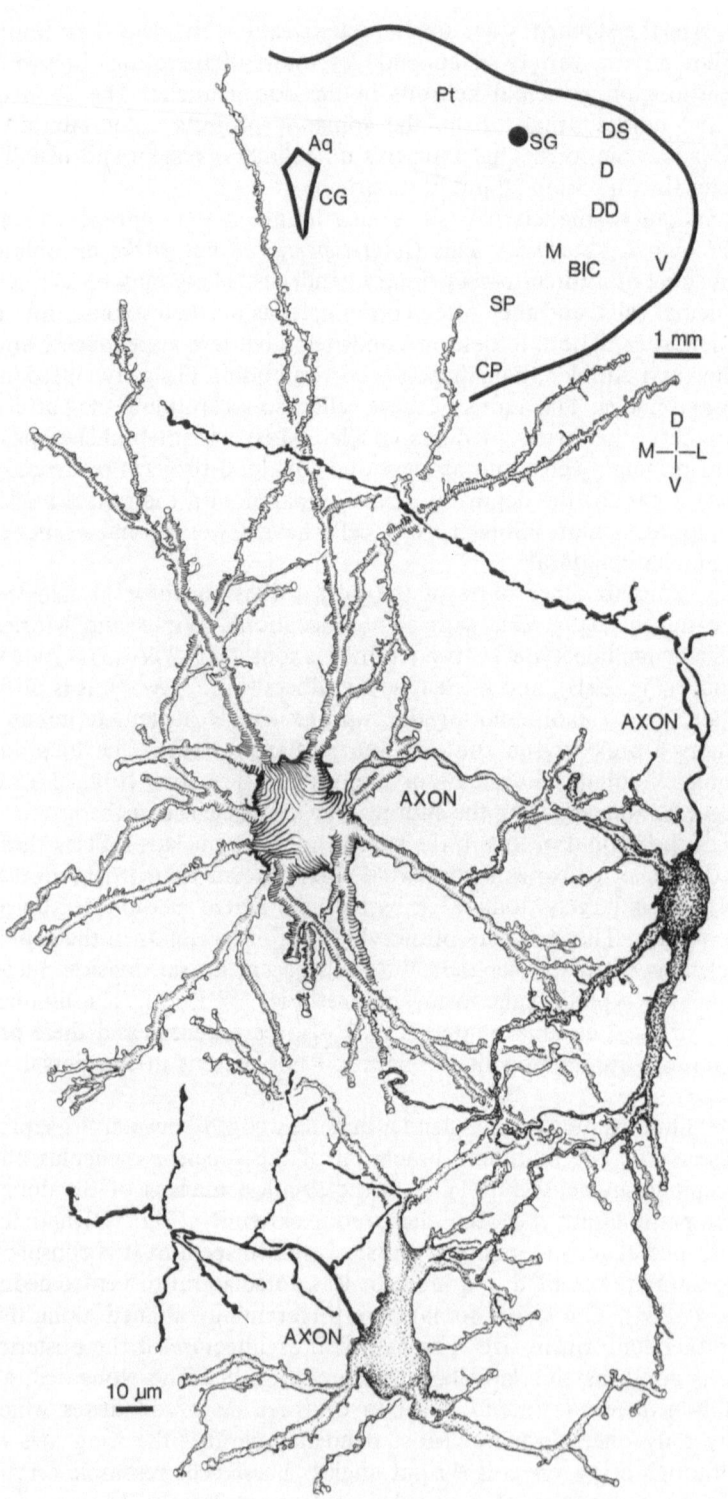

predominantly smooth except for occasional heterogeneous appendages or small aggregates of spines. Lengthy spine-free intervals often have a rough, almost corrugated texture and many swellings and constrictions (Fig. 29B). The axon of this cell type usually originates from the dorsal part of the soma or a proximal dendrite and projects dorsomedially, apparently without, or with only a few, collaterals. The axon is 2–3 μm thick and travels in fascicles with those of other large neurons.

A second type of neuron with a smaller, *stellate dendritic field* is also found in the posterior limitans nucleus. Some of these cells may give rise to axons with local endings and comprise a network of Golgi type II cells (Winer and Morest 1984). The latter class of cells is less common than the poorly branched elongate cells, and the locally arising axonal plexus is less developed than in other parts of the dorsal division.

The course of principal cell axons and their contribution to the dual fiber systems in the posterior limitans nucleus are shown in Fig. 30. The neuropil is dominated by two orthogonal fiber systems. First, the geniculofugal axons of posterior limitans nucleus principal cells course dorsomedially, presumably to leave the thalamus (Fig. 30A). The second fiber system, consisting of axons terminating in and/or passing through the posterior limitans nucleus, is superimposed on the first group (Fig. 30B). Conspicuous here are large numbers of type VIa and type VIb axons coursing in fascicles (Fig. 30B: *above and beneath the stippled soma*), and parts of type VII fibers (Fig. 30B: *5*). Occasionally, a few slender collaterals are observed on what may be principal cell axons (Fig. 30A: *1*). The intrinsic axons in the suprageniculate nucleus are rather less common and robust than in the dorsal nuclei. The texture of the axonal plexus is coarse and somewhat irregular. Dominated by fibers *de passage* and lacking axons with robust intrinsic collaterals, it has a much more "open" texture than the other nuclei of the dorsal division.

3.9 Cortical Connections of the Dorsal Division

Each nucleus of the dorsal division (with the possible exception of the posterior limitans nucleus) has a pattern of overlap and specificity in the organization of its cortical projections. Thus the deep dorsal nucleus (Fig. 31: section *28*) contains neurons labeled with retrogradely transported horseradish peroxidase after injection in cortical area AII (Fig. 5). Labeled cells are also found in the deep dorsal nucleus after injections in the temporal cortex (Fig. 33: section *13, upper right*) and the ventroposterior auditory field (Fig. 34: section *66*).

The cortical projection of the dorsal nucleus also includes, besides AII (Fig. 31: section *28*), the insular cortex (Fig. 32: section *14*), the temporal cortex

◁ **Fig. 26.** The cell types in the suprageniculate nucleus of the dorsal division. The large, irregularly stellate principal neuron has medium-sized radiating dendrites with characteristic, tridentlike terminal branching patterns. The thick axon is without collaterals. Adjacent to this neuron are two Golgi type II cells conspicuous for their thick primary dendrites, small dendritic fields, and locally arborizing axonal collateral systems. (From Winer and Morest 1983a) Rapid Golgi method, 25-day-old cat; planachromat, N.A. 1.32, × 1250

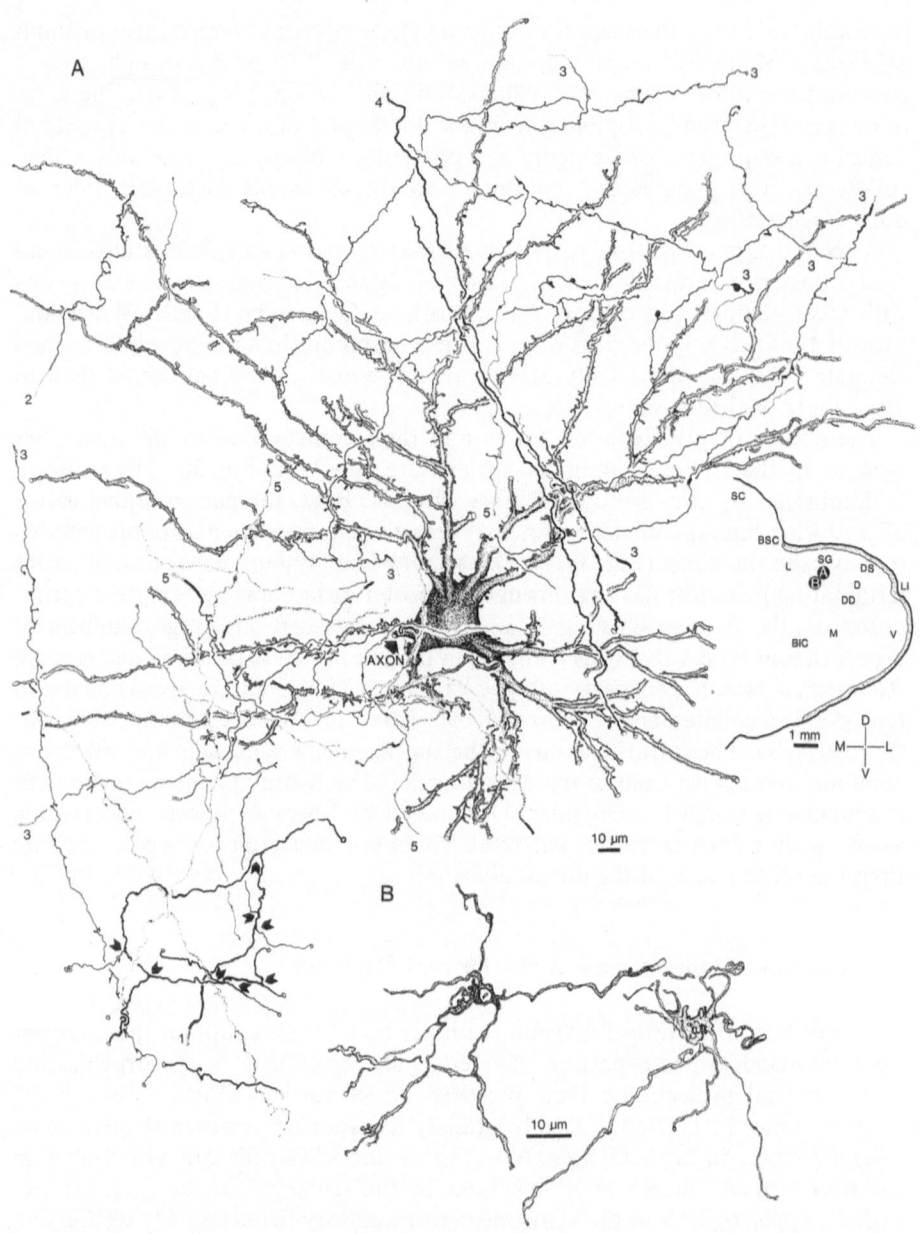

Fig. 27 A, B. A suprageniculate nucleus principal cell and the nearby axonal plexus. **A** Axons adjacent to this cell include type I fibers (*3*), type II axons (*stippled outline, lower left*), fragments of type III and V terminals (*2*), type VIb axons (*5*), and type VIa fibers (*upper right*). Axons from Golgi type II cells (*1*) and their collaterals (*six-sided arrowheads*) are also present. **B** Group IV grumous endings in the suprageniculate nucleus at higher power. Although somewhat smaller and simpler in structure than comparable endings in the dorsal nuclei, they have a similar central body and lacunae (see Fig. 23). (Modified from Winer and Morest 1984) Rapid Golgi method, 41-day-old cat; planapochromat, N.A. 1.32, × 1250

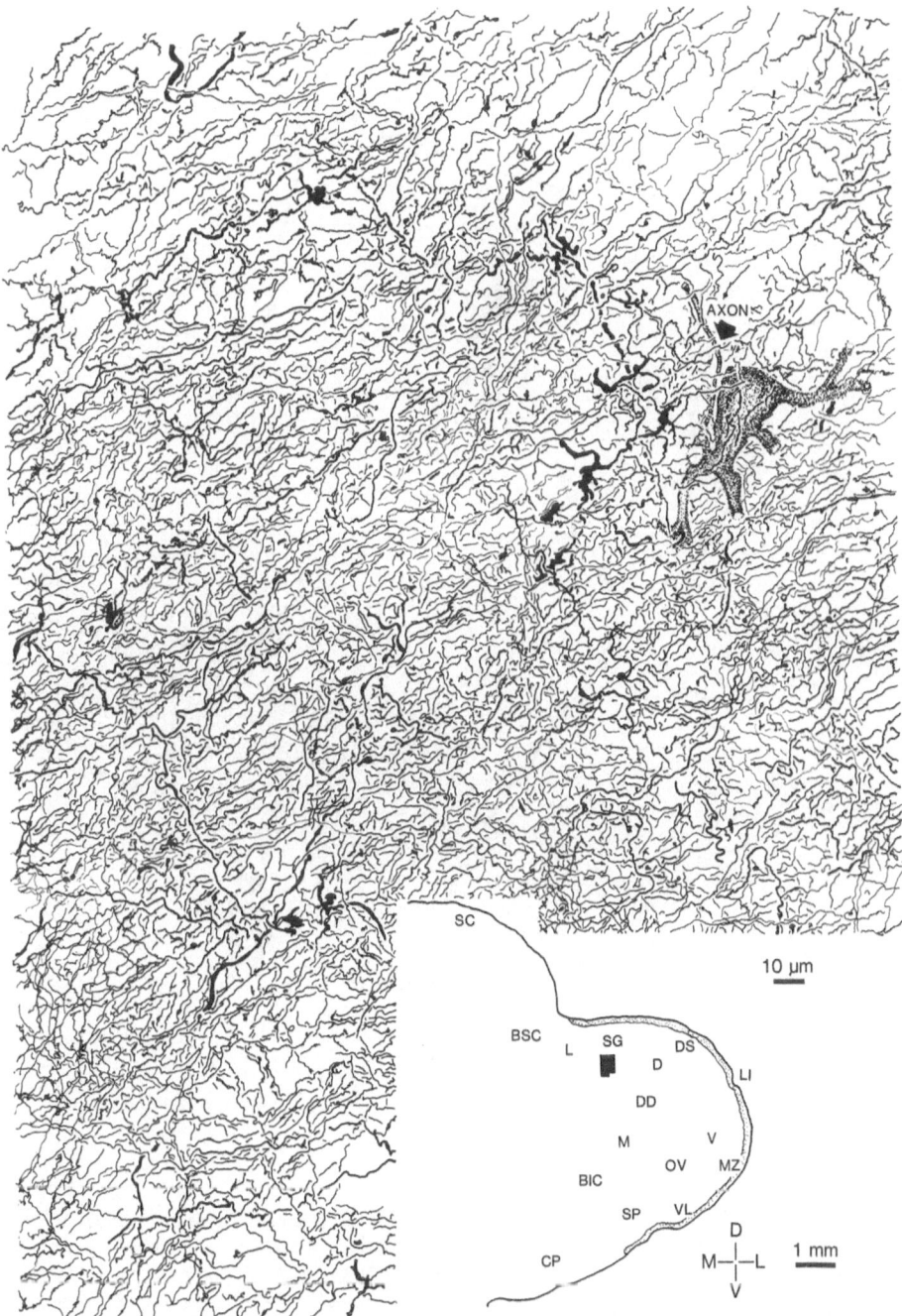

Fig. 28. The axonal plexus of the suprageniculate nucleus. The texture of this neuropil has a less regular, more fasciculated appearance than similar 100-μm-thick sections from the dorsal nuclei (see Fig. 24). Most fibers are of parabrachial origin (Fig. 25) and have an oblique, nearly horizontal trajectory. Brachial fibers running from ventromedial-to-dorsolateral, some type IV endings (*black outlines*), and the fine collaterals of Golgi type II axons are evident. Some of the areas of low staining density probably contain unimpregnated somata. (From Winer and Morest 1984) Rapid Golgi method, 41-day-old cat; planapochromat, N.A. 1.32, ×1250

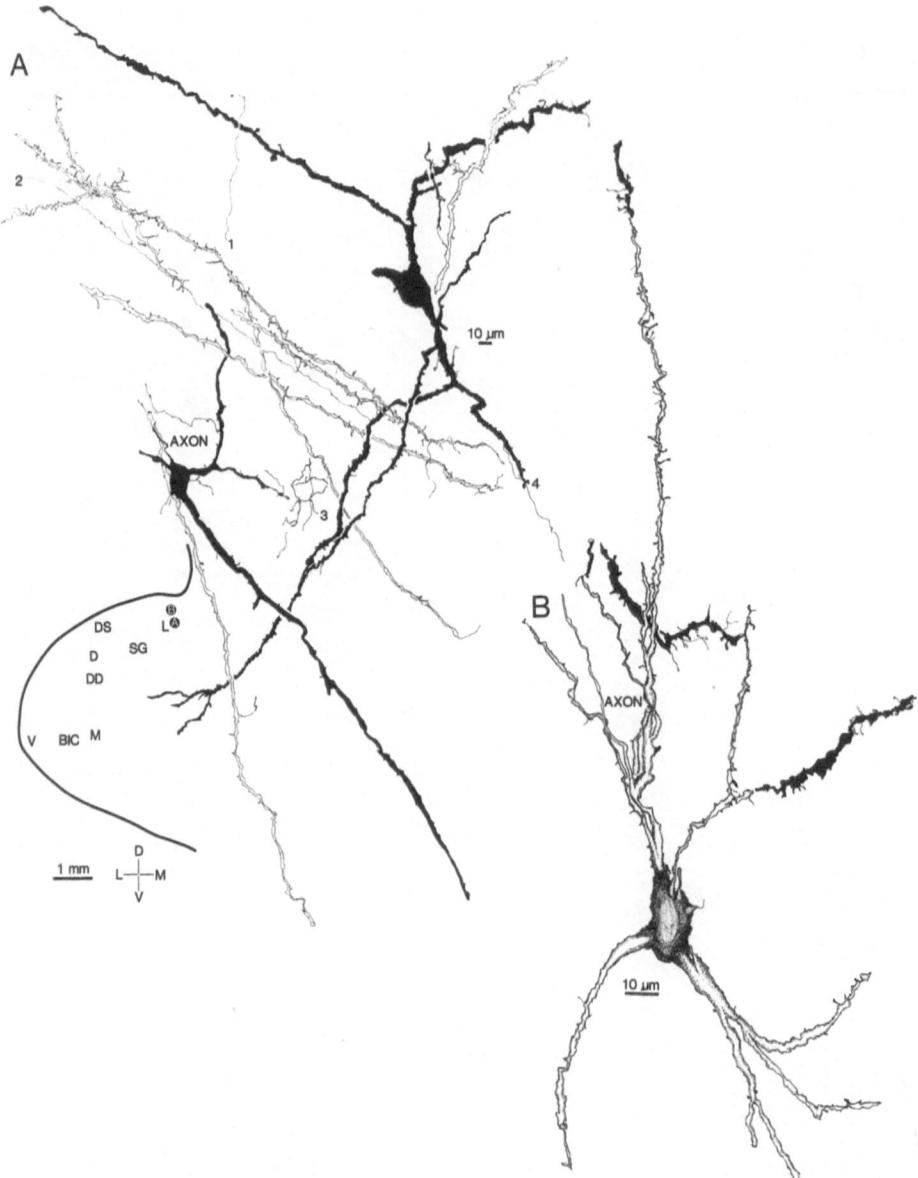

Fig. 29 A, B. (Above) The elongated, fusiform cell of the posterior limitans nucleus of the dorsal division. **A** The pear-shaped somata have polarized dendrites which conform to the long axis of the nucleus; only a few dendrites cross obliquely. Rapid Golgi method, 23-day-old cat; planachromat, N.A. 0.75, × 500. **B** A fusiform cell at higher power showing the characteristic delicate dendritic appendages and the initial segment of the axon (*open outline*). Note also the somatic spines. *Solid black,* adjacent principal cell dendrites. Rapid Golgi method, 23-day-old cat; planachromat, N.A. 1.32, × 1250

Fig. 30 A, B. (To the right) The two systems of axons and the axonal plexus in the posterior ▷ limitans nucleus. **A** The dual fiber systems. The large, unbranched fibers passing dorsomedially belong to principal cells. A second group of fibers, afferent to the posterior limitans nucleus, is formed by type I axons (*4*), type III axons (*2*), type V axons (*6*), type VIa axons (*1, upper left*), and type VIb axons (*lower right;* see Fig. 25). Rapid Golgi method, 23-day-old cat;

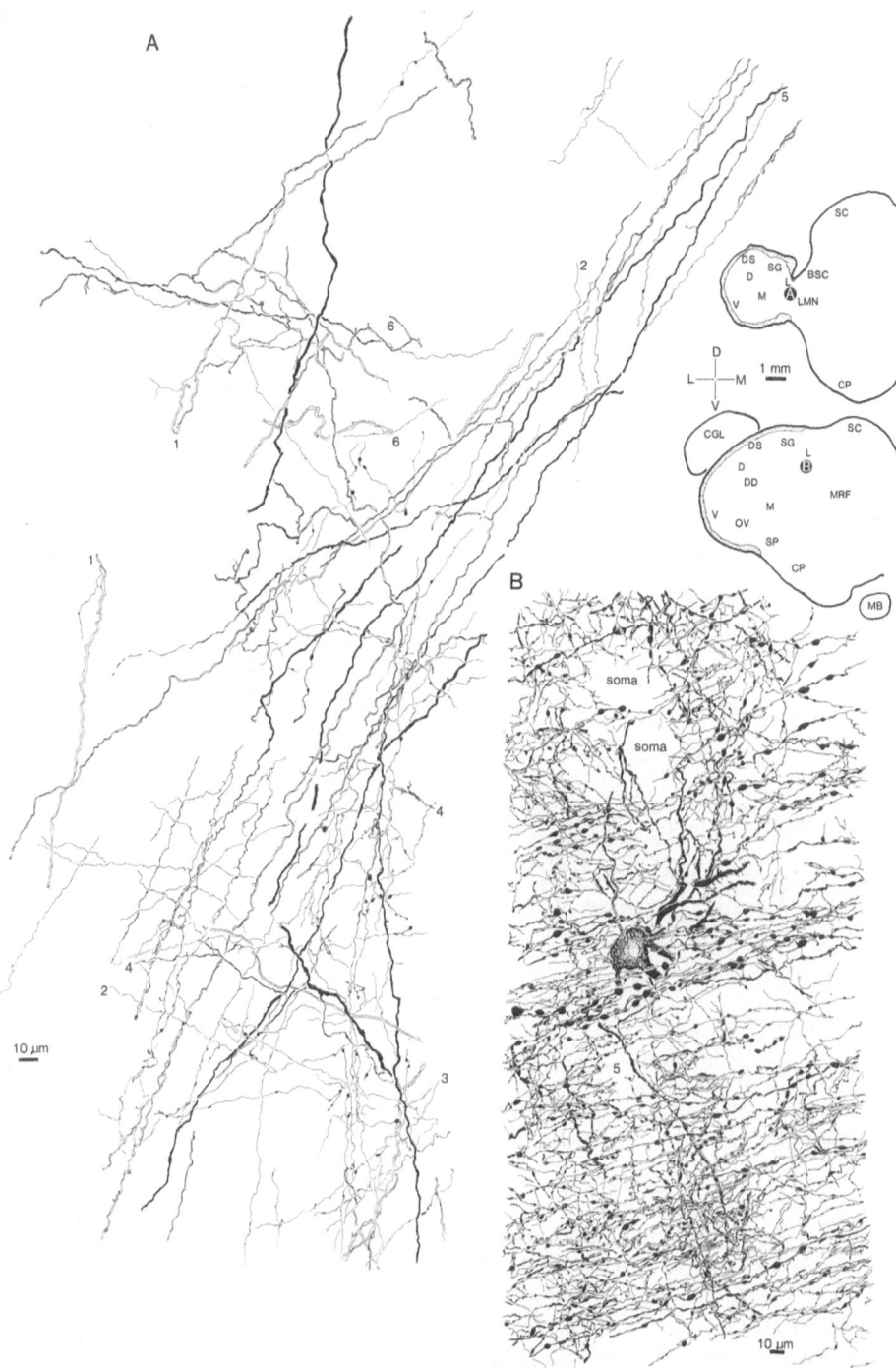

planapochromat, N.A. 1.32, ×1250. **B** The loosely organized, fasciculated posterior limitans nucleus neuropil in a 100-μm-thick section. Few axons attributable to Golgi type II cells are present, and the open areas (*5*) probably contain somata. A principal cell soma is impregnated (*fine stipple*). (Modified from Winer and Morest 1984) Rapid Golgi method, 18-day-old cat; planapochromat, N.A. 1.32, ×1250

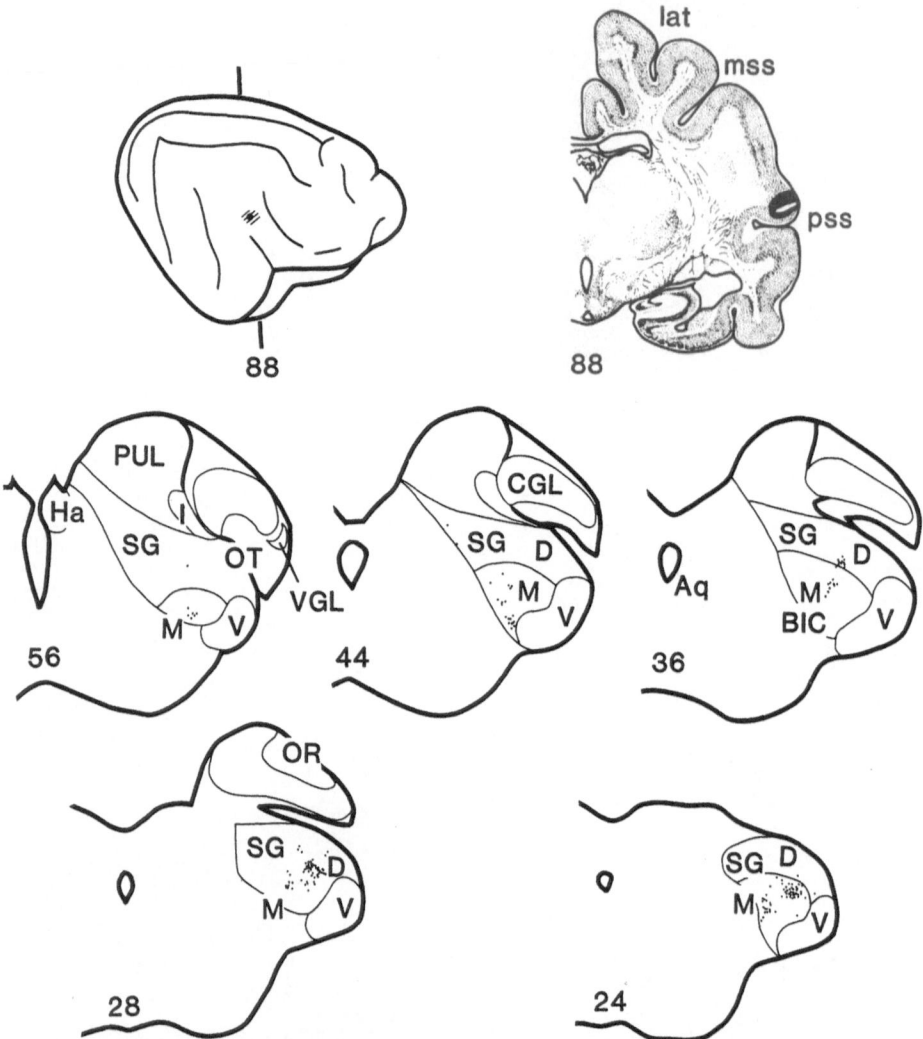

Fig. 31. The distribution of labeled neurons in the dorsal division after an injection of 0.25 μl of horseradish peroxidase in the second auditory cortical area (AII). The greatest number of labeled cells are concentrated in the deep dorsal nucleus and the medial division; a few scattered neurons lie in the suprageniculate nucleus. Adult cat. (Redrawn from Winer et al. 1977)

(Fig. 33: section *6*), and the ventroposterior auditory field (Fig. 34: section *73*). The heaviest projection is to the ventroposterior field.

The superficial dorsal nucleus has a more restricted target. These axons terminate in insular cortex (Fig. 32: section *30*), in the temporal cortex (Fig. 33: section *18*), and in the ventroposterior field (Fig. 34: sections *66, 74*). The projection of this nucleus includes non-primary auditory cortex (Fig. 5) with the probable exception of AII.

Suprageniculate nucleus cells continue the pattern of partial specificity and partial overlap in the cortical projections of dorsal division nuclei. Their primary target is the insular cortex rostral to the pseudosylvian sulcus (Fig. 32). While

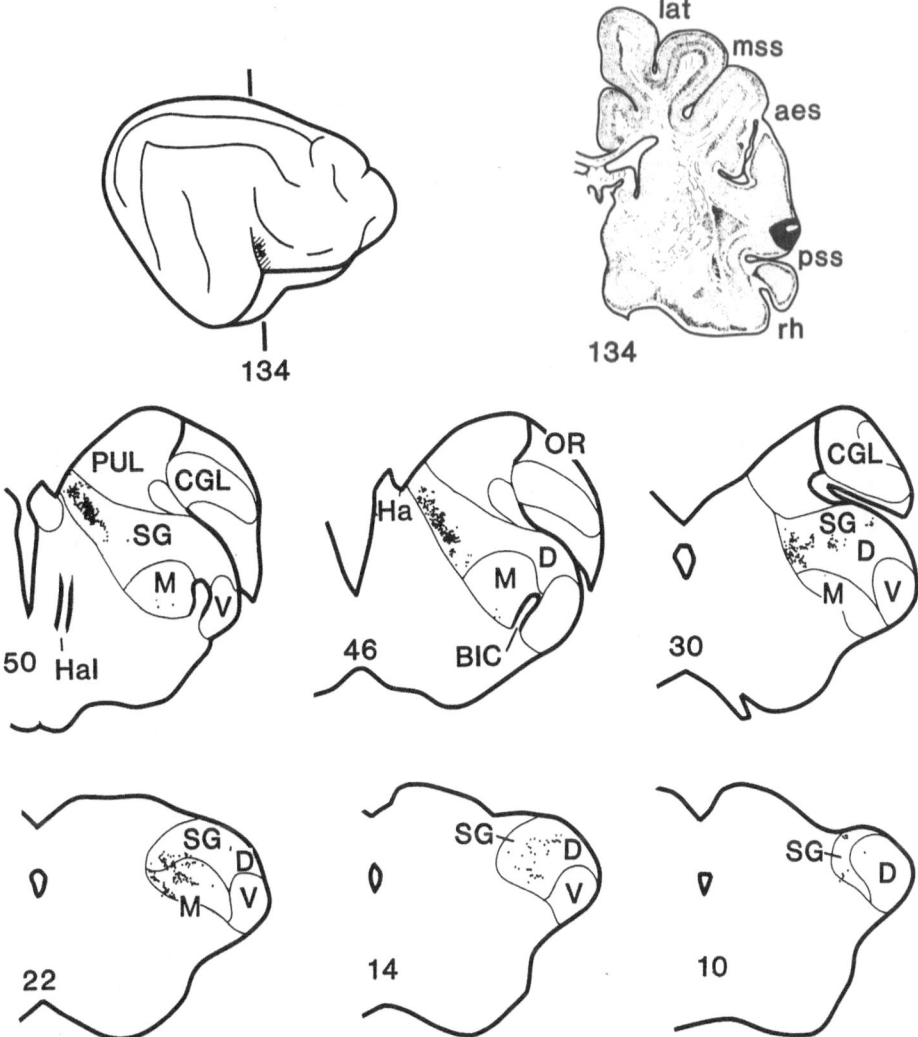

Fig. 32. The distribution of labeled cells after two injections of 0.1-μl each of horseradish peroxidase in the insular cortex. Most labeled cells lie in the medial limb of the suprageniculate nucleus. Some are found in the superficial dorsal and dorsal nuclei as well as the medial division. Adult cat. (Redrawn from Winer et al. 1977)

this projection is heavy (Fig. 32: sections *30, 16*), at least some suprageniculate neurons (or their axonal branches) project to the temporal cortex behind the pseudosylvian sulcus (Fig. 33: section *18*). A much lighter projection to the ventroposterior field is also found (Fig. 34: section *73*). In each of the foregoing experiments (Figs. 31–34), as well as in experiments involving primary auditory cortex (Fig. 13), some labeled neurons are scattered in the medial division. Finally, the presence of labeled cells in the ventrolateral nucleus (Fig. 34: section *90*) after an injection in the ventroposterior field argues that this nucleus should be included, on connectional grounds, with the dorsal division. This is also supported by the morphology of its neurons (Figs. 1A, 2B), which are primarily

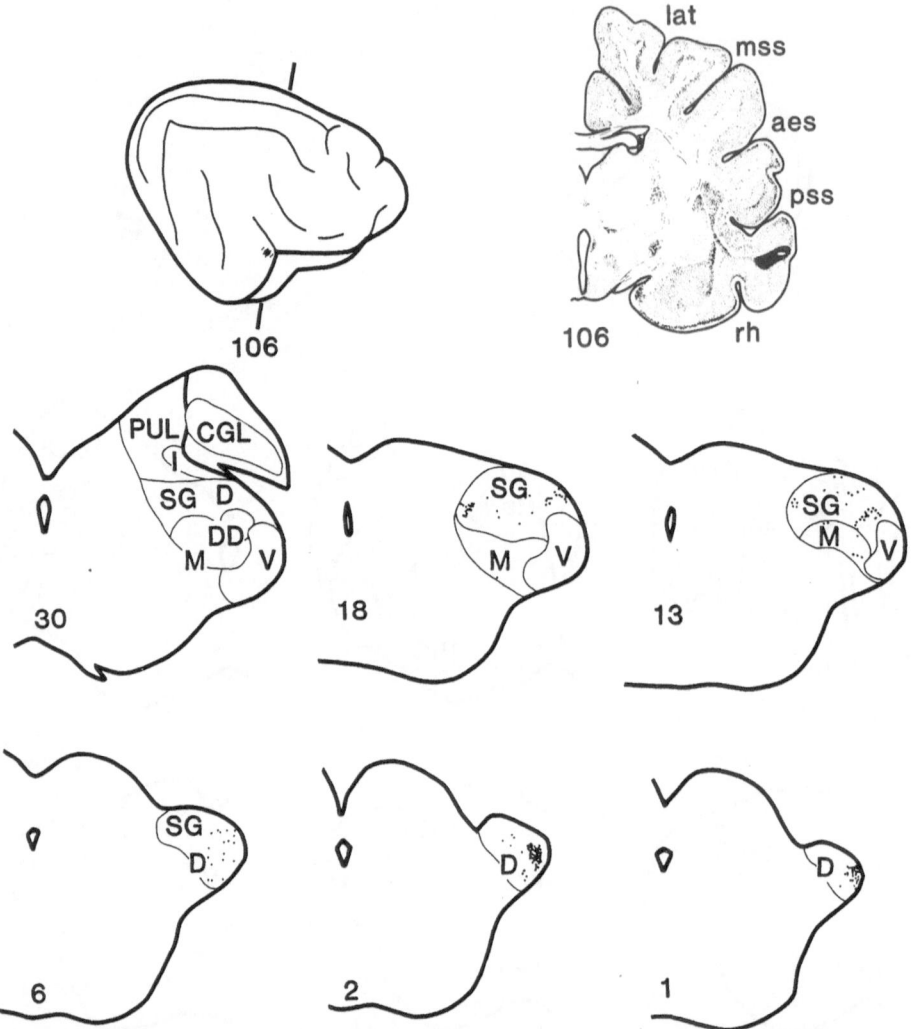

Fig. 33. The distribution of labeled neurons after an injection of 0.3 µl of horseradish peroxidase in the temporal cortex caudal to the pseudosylvian sulcus. Labeled cells lie in the caudal dorsal, superficial dorsal, dorsal, and deep dorsal nuclei. A few labeled cells are found in the medial division. Adult cat. (Redrawn from Winer et al. 1977)

stellate in structure (and thus allied with the dorsal division). Geographically, however, they lie within the ventral division.

3.10 Neuronal and Axonal Architecture of the Medial Division

The organization of the medial division contrasts sharply with that of the ventral and dorsal divisions. It consists of morphologically heterogeneous neurons which cannot (in the cat) be differentiated into particular nuclei. It alone of the subdivisions of the medial geniculate body has a pattern of neuropil where

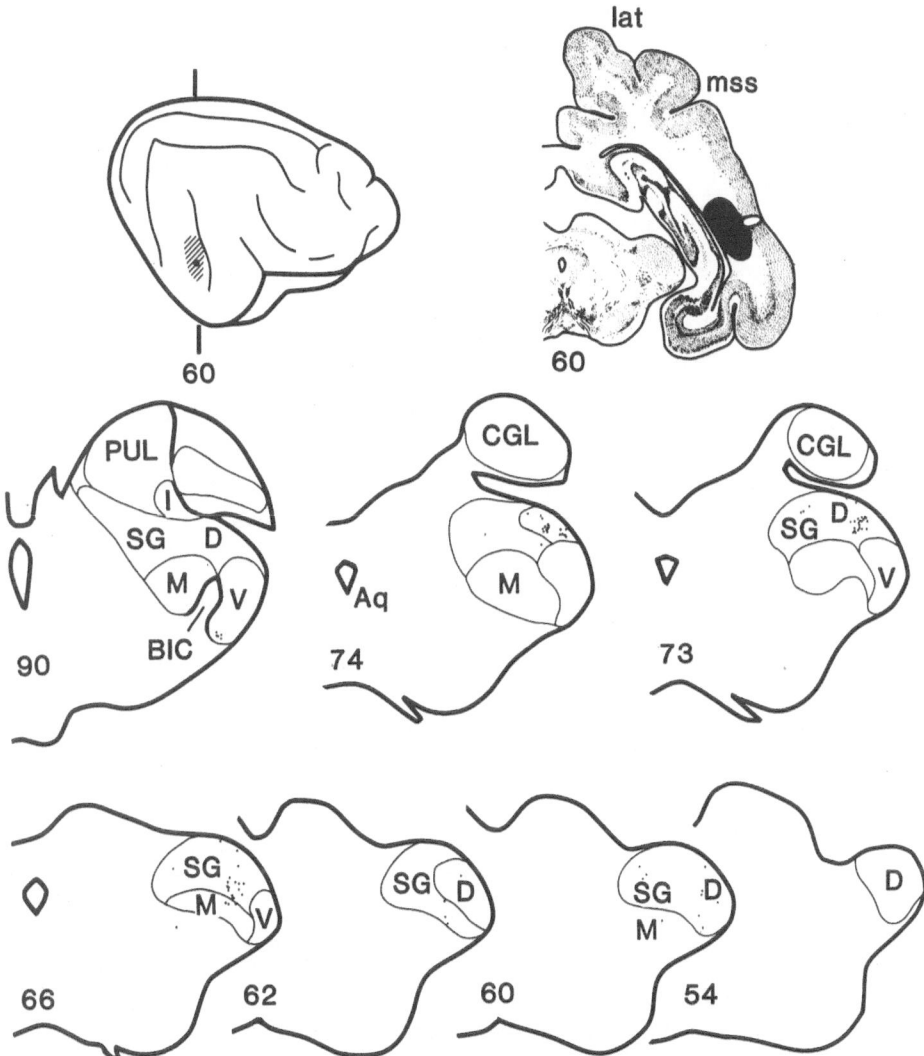

Fig. 34. The distribution of labeled cells after an injection of 0.3 µl of horseradish peroxidase in the ventroposterior auditory field of the posterior ectosylvian gyrus. Labeled cells are found in the superficial dorsal, dorsal, and deep dorsal nuclei; injections of the posterior ectosylvian gyrus always label neurons in the ventrolateral nucleus (section *90, beneath V;* see text and Table 2). Some medial division cells are also labeled. Adult cat. (Redrawn from Winer et al. 1977)

the axons of Golgi type II cells appear to play a limited role. It also has a pattern of cortical projections different from the other divisions of the medial geniculate body.

In Golgi material (Fig. 1 A) the medial division is apparent for its large neurons and their relatively low density (Fig. 2 B). The medial division as such is not present in the most caudal parts of the medial geniculate body (Figs. 2 A, 3 A). At this level only the dorsal nuclei and scattered, medium-sized cells of the interstitial nuclei of the brachium of the inferior colliculus are present (Winer

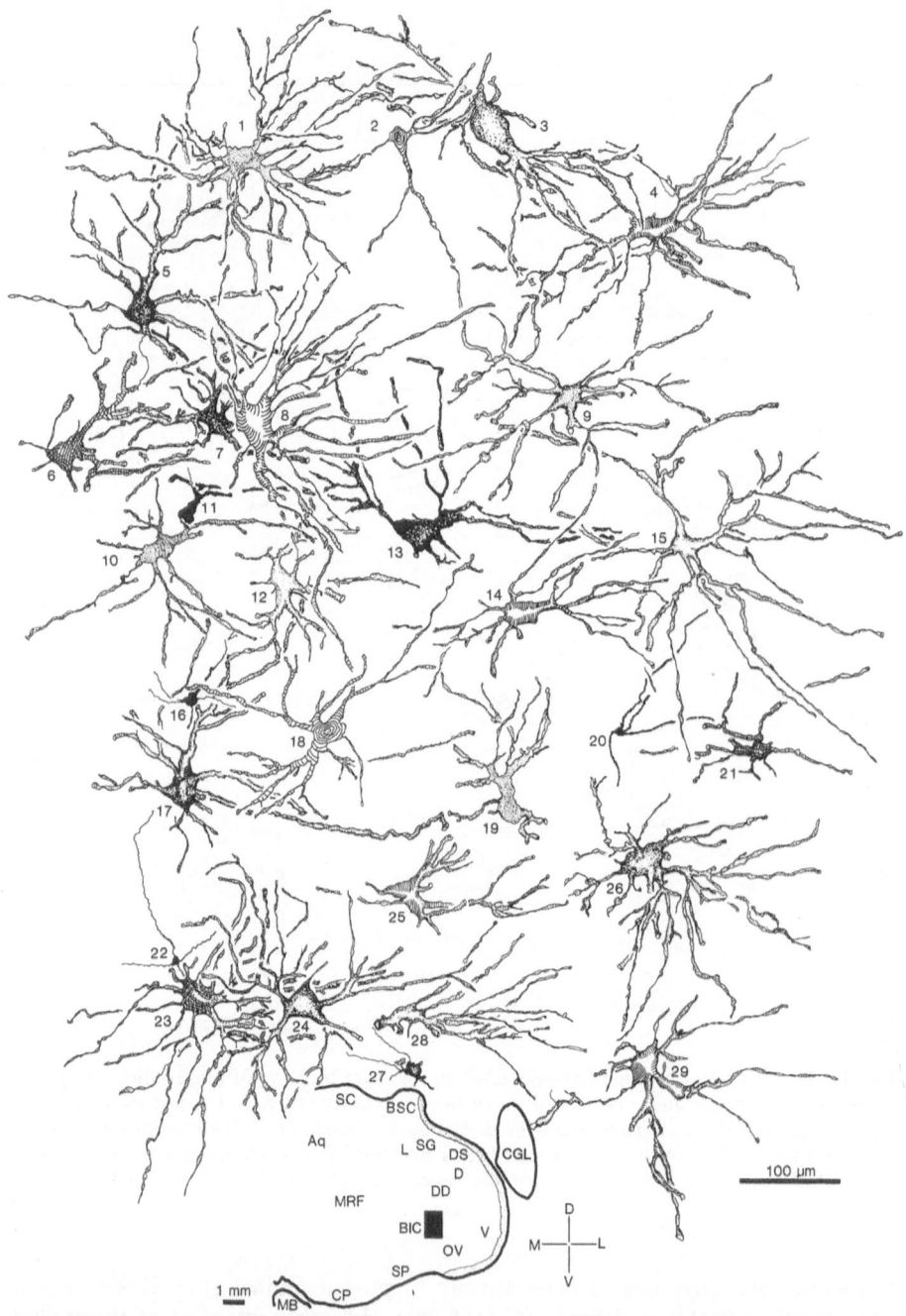

Fig. 35. The morphology of medial division neurons drawn and superimposed from four serial, 100-μm-thick sections. The heterogeneity of these cells is apparent, and includes stellate neurons (e.g., *8, 9, 15, 18*), bushy cells (e.g., *3, 6, 25, 28*), neurons with wide-field dendrites (e.g., *4, 10, 14, 17*), magnocellular neurons (e.g., *1, 13, 23, 24*), and smaller stellate cells (e.g., *11, 20, 21, 27*). (From Winer and Morest 1983b) Golgi-Cox method, adult cat; planachromat, N.A. 0.35, ×200

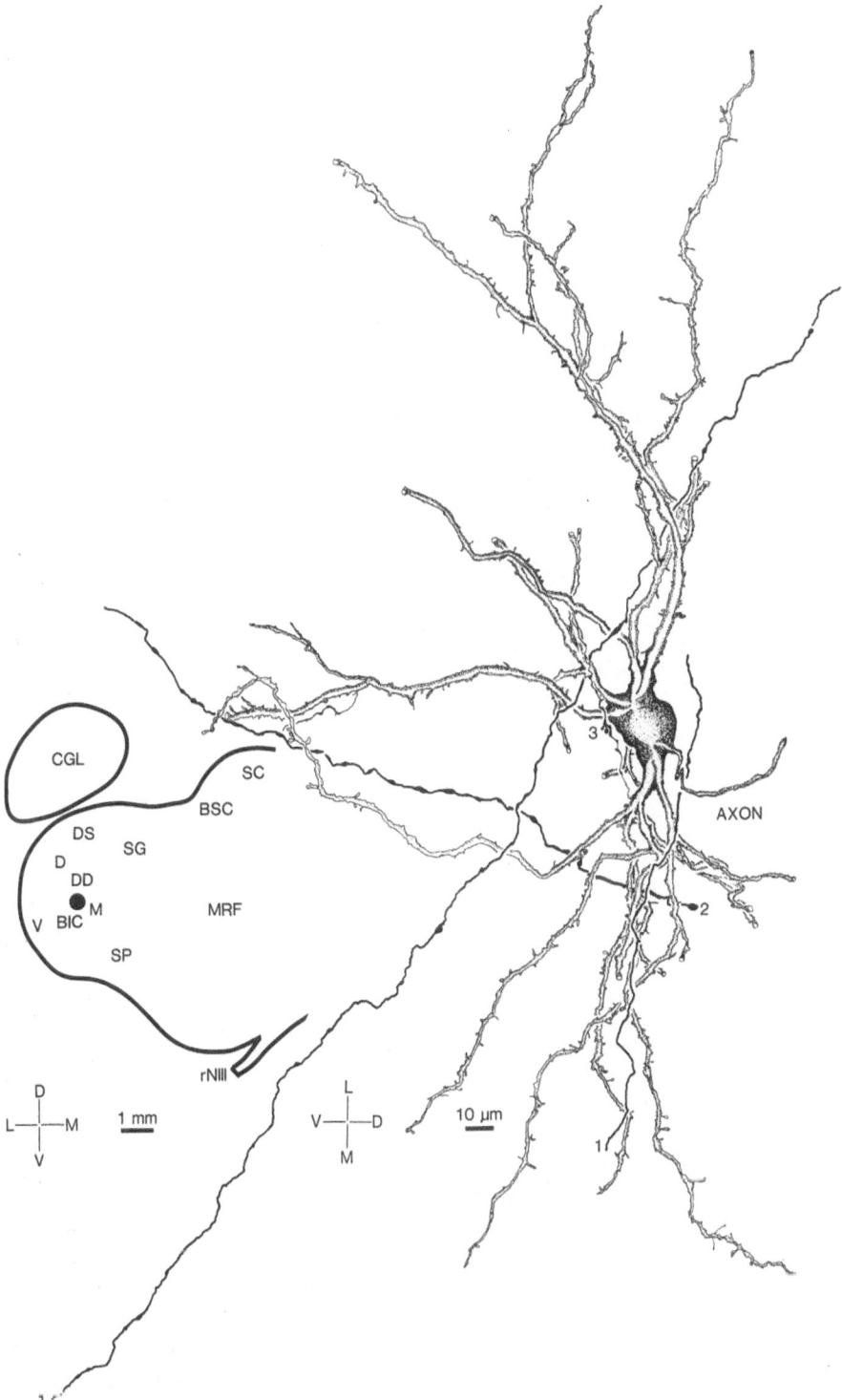

Fig. 36. A medial division cell with wide-field dendrites which parallel and span the brachial fibers. The slender dendrites intertwine with each other at their distal portions. Delicate appendages are common on the central parts of dendrites, and the axon is medium-sized and lacks a collateral system. Various preterminal axons also occur (*1–3*). Rapid Golgi method, 29-day-old cat; semi-apochromat, N.A. 1.25, × 1250

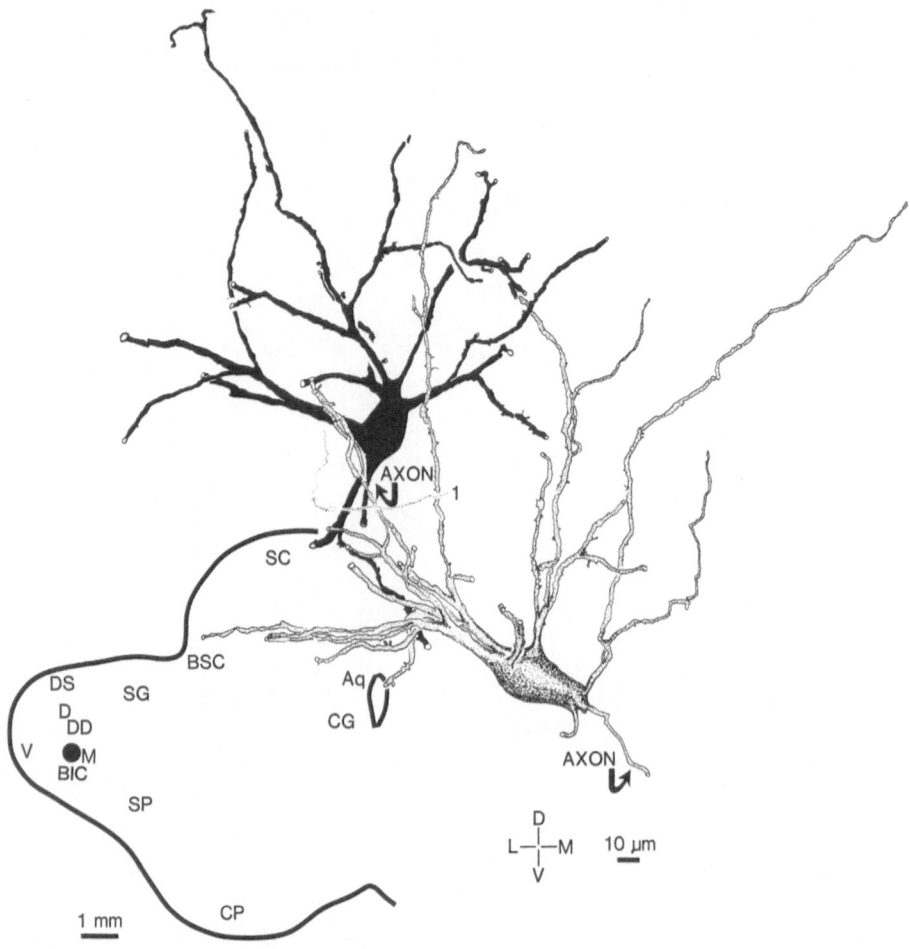

Fig. 37. A large stellate cell (*solid black*) and a bushy neuron (*stippled*) from the medial division. The irregular, rather smooth dendrites of these cells often mingle with each other among the brachial fibers. A thick axon without collaterals issues from the soma. A fine afferent (*1*) is present. Rapid Golgi method, 39-day-old cat; planapochromat, N.A. 1.32, × 1250

and Morest 1984). The full expansion of the brachium marks also the development of the medial division (Fig. 2B). In Nissl material (Figs. 3, 4A) the hallmarks of medial division organization — large cells with low packing density and lacking preferential somatic orientation (Fig. 4I) — are clear. More rostrally (Fig. 2C, D) the dispersion of cells is accentuated by the complex, variegated fiber architecture.

Golgi material confirms that the medial division is a mixture of cell types (Fig. 35) and that no single variety is dominant. While these neurons are larger than those of other divisions of the medial geniculate body (Morest 1964), many intermediate-sized and small cells are present. *Medium-sized neurons* may have wide (Fig. 35: *2, 10*) or bushy (Fig. 35: *25, 28*) dendritic domains. Intermediate and large multipolar cells predominate. The former have broad or wide-field dendritic arbors (Fig. 35: *4, 14, 15*), while the magnocellular neurons

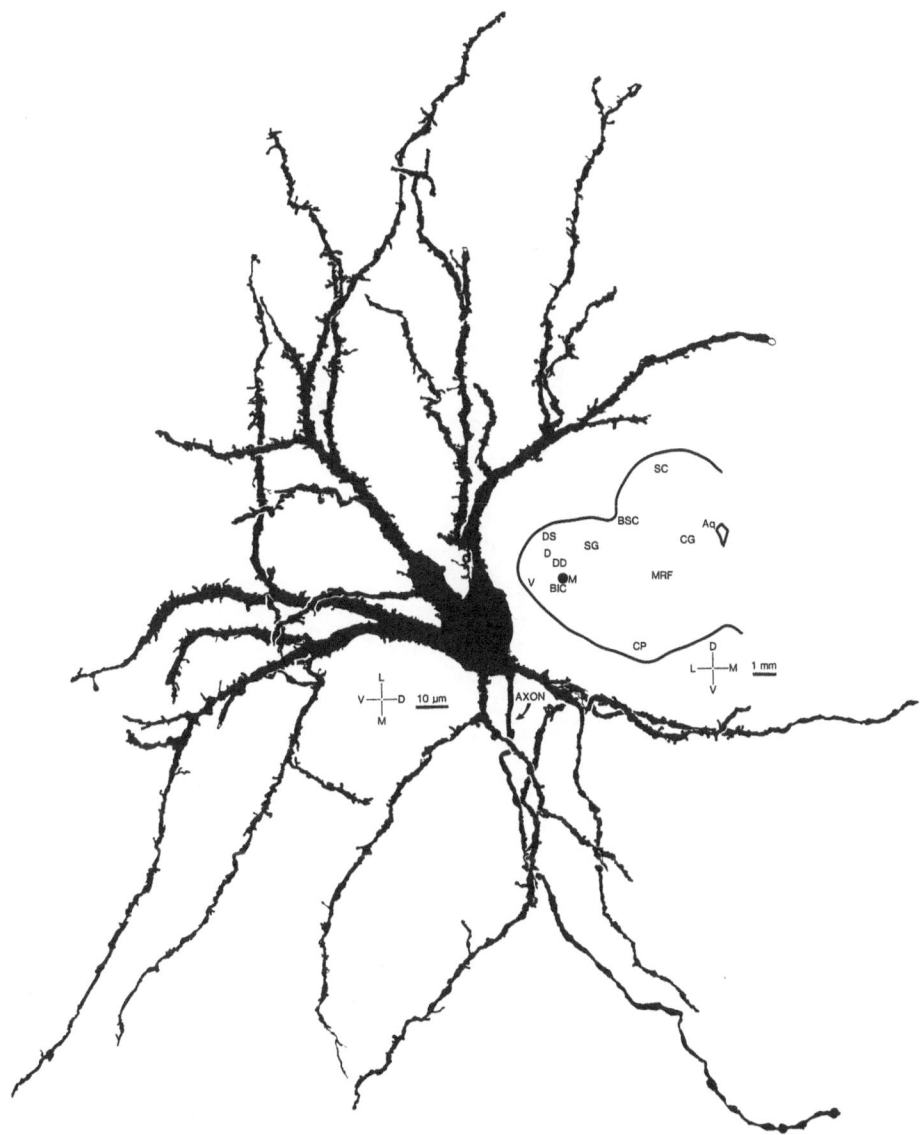

Fig. 38. A medium-sized neuron with irregularly stellate dendrites from the medial division. This cell has morphologically heterogeneous dendritic spines, including short appendages, long and thin varieties with or without pedunculated endings, and other types. Some of the proximal dendrites are very thick. The axon is 2-μm thick and free of collaterals but has conspicuous dilatations. Rapid Golgi method, 23-day-old cat; planachromat, N.A. 1.32, × 1250

(Fig. 35: *1*, *26*) have somewhat smaller, modified radiate dendritic fields and a stellate branching pattern.

Neurons with wide-field dendritic patterns are common in the medial division (Fig. 36). This cell has an oval soma whose long axis and dendrites are parallel to brachial fibers (Fig. 36: *1*). Three or four major dendrites emerge from the medial and lateral somatic poles and extend laterally for up to 250 μm before

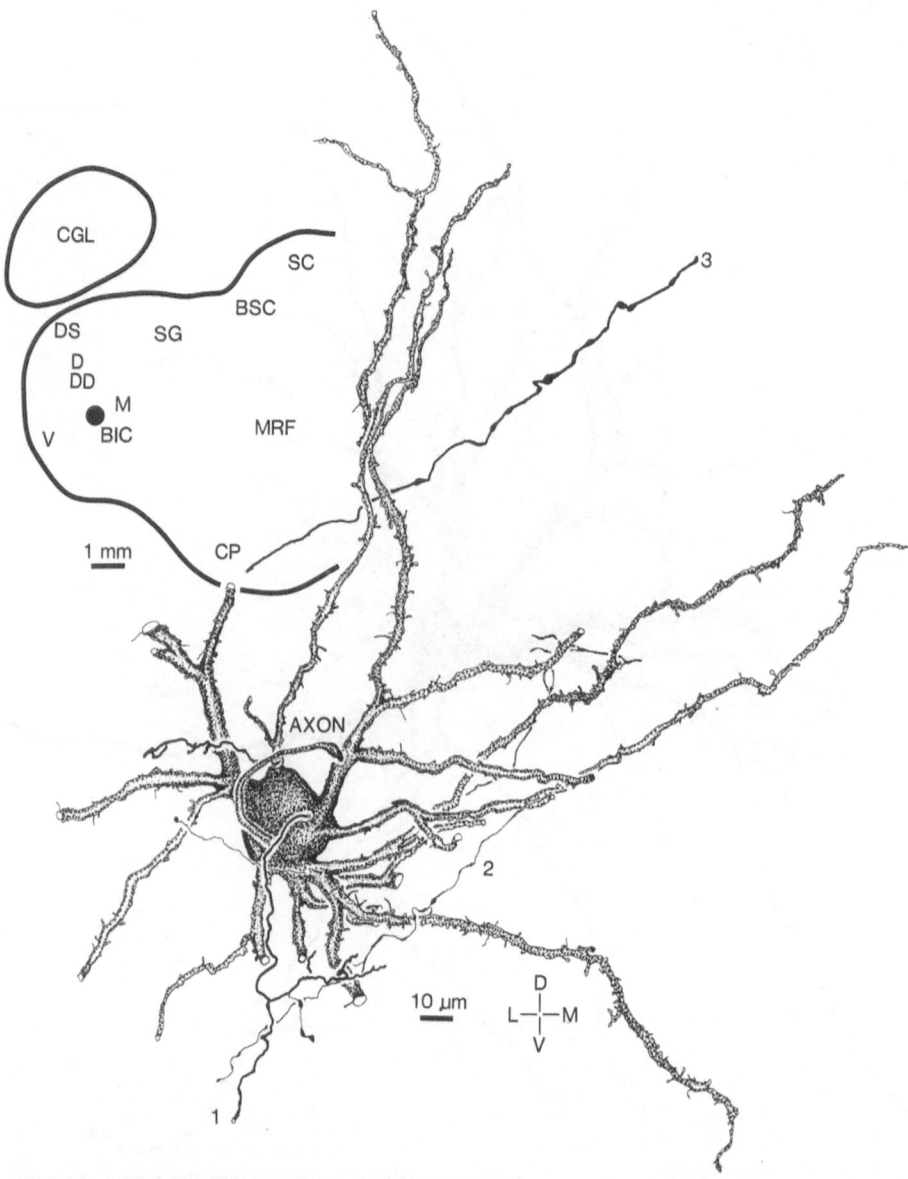

Fig. 39. A magnocellular neuron from the medial division. This massive, round perikaryon has surprisingly slender dendrites with an irregularly stellate branching pattern. The bulk of the many types of dendritic spines are found upon the central dendritic shaft. The thick axon has no collaterals and projects toward the crossing brachial fibers (see Fig. 43). Afferent axons form thick, perisomatic branches (*1*) and thinner, peridendritic fibers (*2*). Rapid Golgi method, 29-day-old cat; semi-apochromat, N.A. 1.25, × 1250

ending in the plane of the section. These dendrites often wind about each other, especially at their ends, and may branch obliquely. A few dendrites project ventrally and parallel other, possibly descending, axons (Fig. 36: *2*). The dendritic appendages are short and spiculated, though longer and pedunculated spines also occur. The axon usually emerges from the dorsal part of the soma,

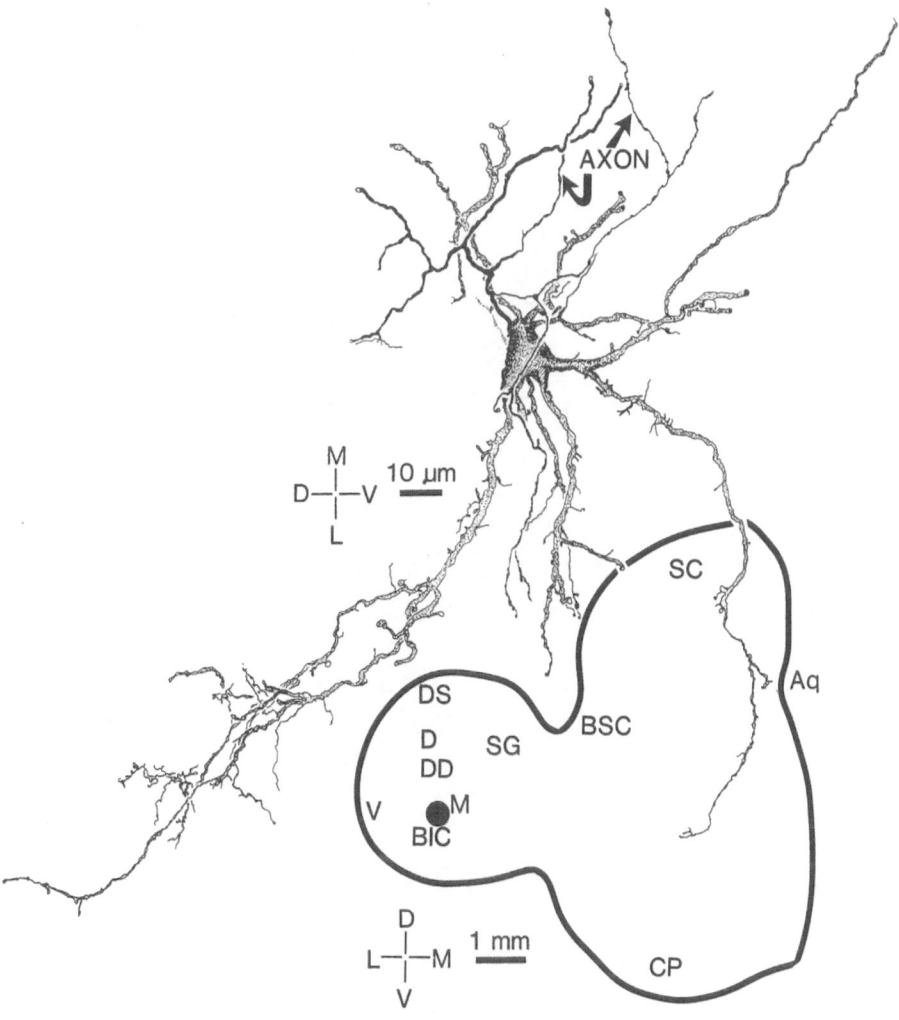

Fig. 40. A Golgi type II cell from the medial division. This neuron has a small soma, thin dendrites with long, delicate appendages, and an axon (*arrows*) which ramifies in the vicinity of the cell's dendrites. The distal dendrites are thin, and terminal axonal collaterals are often extremely slender. Rapid Golgi method, 23-day-old cat; planapochromat, N.A. 1.32, × 1250

is about 2 μm thick, and is apparently without collaterals. It acquires its myelin sheath within 50 μm of its origin.

Somewhat larger than the wide-field neurons are *cells with bushy dendrites* (Fig. 37: *stippled outline*) or *modified stellate dendrites* (Fig. 37: *black outline*). These neurons commingle with other medial division cell types. The bushy cell has an elongated soma and four-to-six thick dendrites which branch several times near the cell. The terminal parts of its rather smooth dendrites often form cone-shaped expansions which project outside the section. The modified stellate cells have even fewer dendritic spines and a simpler pattern of dendritic branching. A thin axon, apparently without collaterals, emerges from the elongated 25 × 40 μm soma.

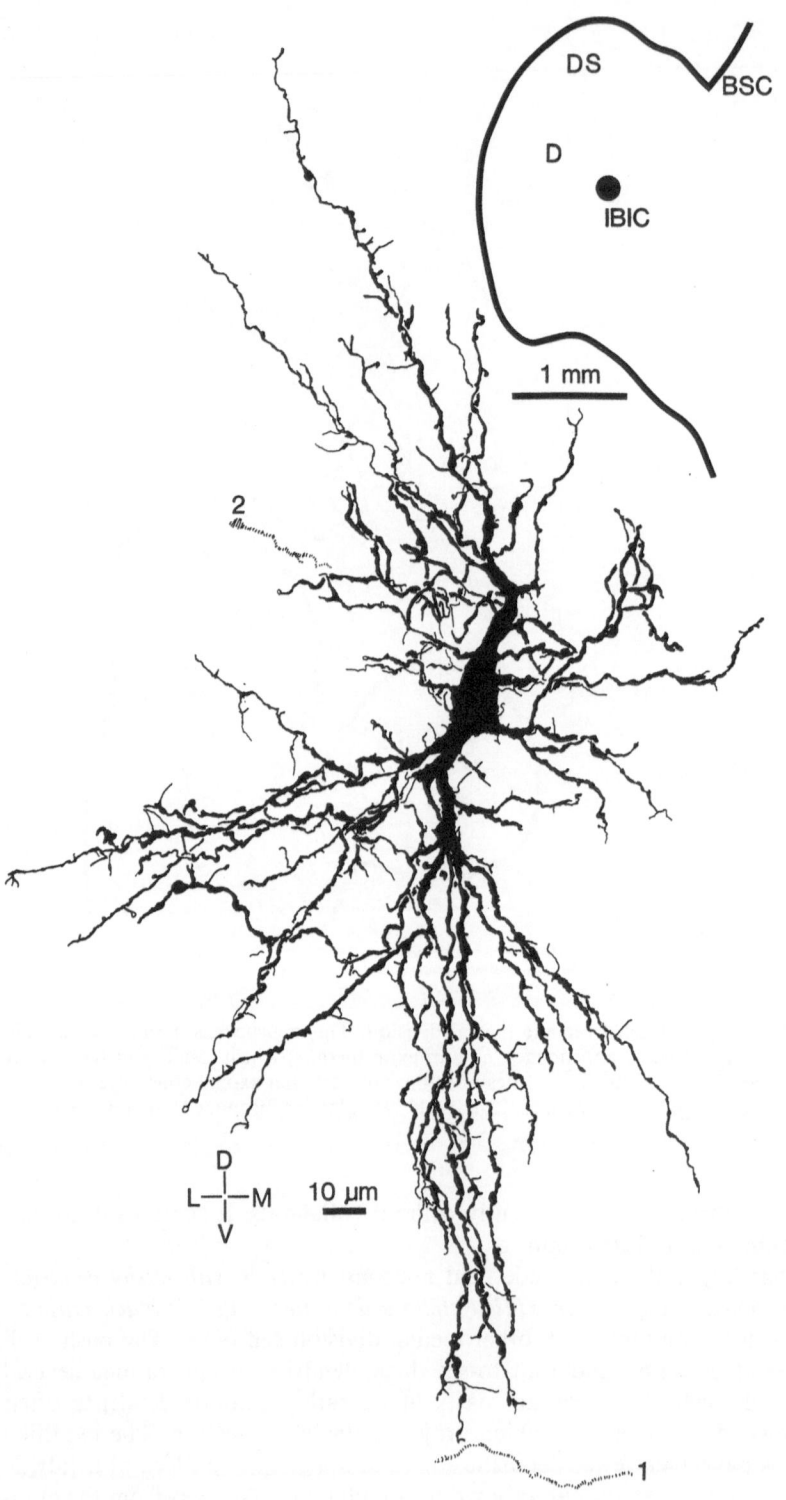

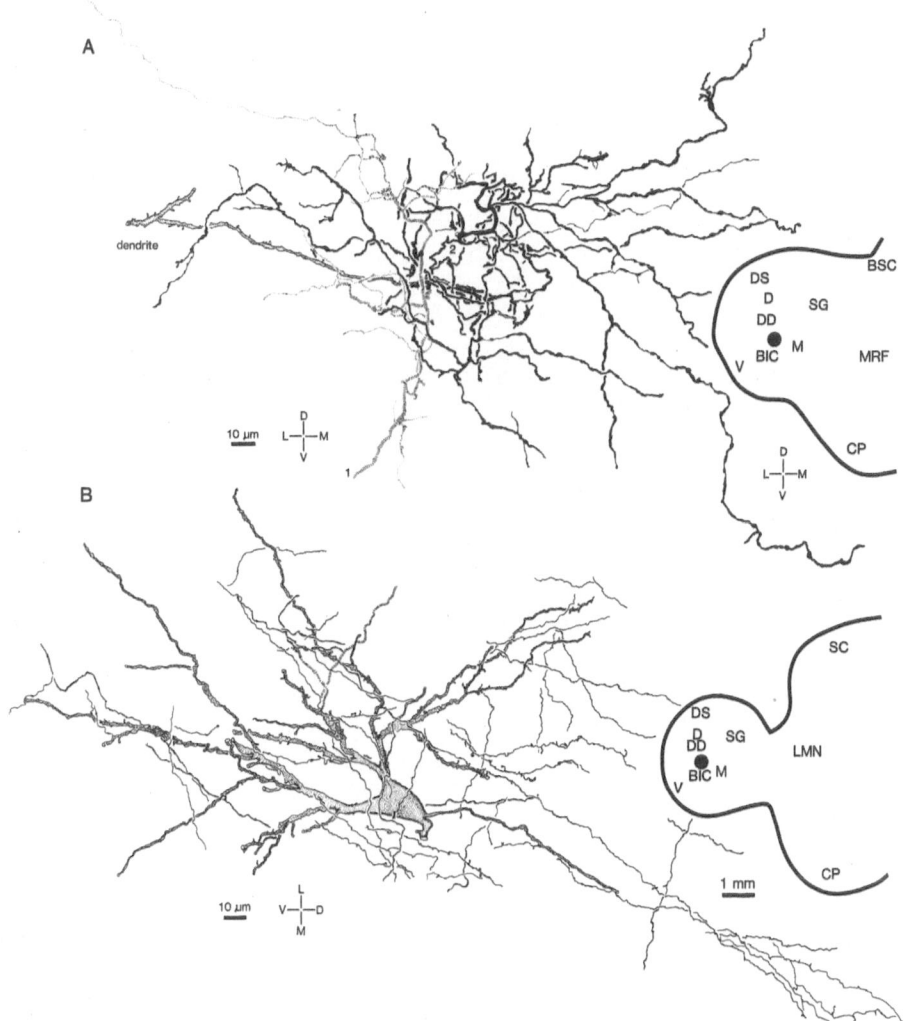

Fig. 42A, B. Some extrinsic axons in the medial division and their terminal plexus. **A** Massive, serpentine endings (*1, 2*) forming gnarled, dense terminal nests about a principal cell dendrite. Rapid Golgi method, 44-day-old cat; planapochromat, N.A. 1.32, × 1250. **B** The plexus of predominantly fine axons afferent to a bushy neuron. (From Winer and Morest 1983b) Rapid Golgi method, 36-day-old cat; planapochromat, N.A. 1.32, × 1250

A *medium-sized cell with radiate dendrites* and a round, 30–35 µm soma is common in the medial division (Fig. 38). The four-to-six very thick dendritic trunks emerge irregularly from the soma, usually branching only once to form Y-shaped junctions, and have more appendages than the cell types described

◁ **Fig. 41.** A neuron from the caudal part of the interstitial nucleus of the brachium of the inferior colliculus. This immature cell has profuse dendritic branchlets and is found among fibers of the caudal brachium just behind the medial division, where few such bushy cells are present (see Fig. 2A, D). At this level, the medial division is not yet developed, and these interstitial nucleus neurons span the brachial fibers (*1, 2;* see also Fig. 43). Rapid Golgi method, 6-day-old cat; planapochromat, N.A. 1.32, × 1250

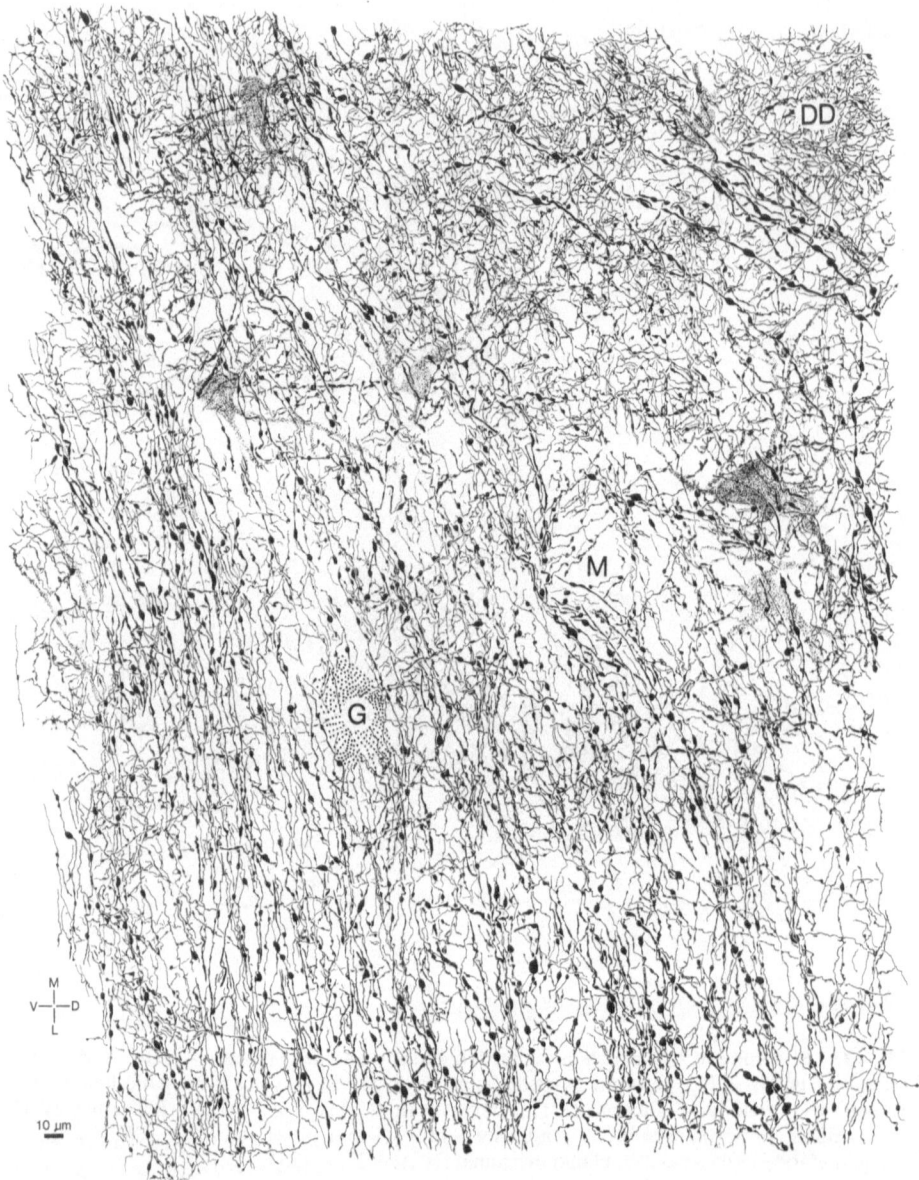

Fig. 43. The axonal neuropil of the medial division (*M*) where it adjoins the deep dorsal nucleus (*DD*). All the well-impregnated axons from a 100-μm-thick section were drawn to show the relation between the fascicles of ascending brachial fibers coursing medio-laterally, the descending fibers running orthogonal to these (see Fig. 42 B), and the fine, heavily stained, concentrated plexus of the adjoining deep dorsal nucleus (see Fig. 24). Various medial division cells are shown (*stippled outlines*) including several glial cells (*G*). (From Winer 1979 and Winer and Morest 1983 b) Rapid Golgi method, 18-day-old cat; semi-apochromat, N.A. 1.25, ×1250

Fig. 44 A–E. Photomicrographs of Nissl-stained neurons in medial geniculate body subdivi- ▷ sions. **A** Principal cells from *pars lateralis* with polarized somata. **B** A small cell (*open arrow*) between two principal neurons in *pars lateralis*. **C** Neurons from the superificial dorsal nucleus. Cells with round perikarya are present, as is a neuron with prominent somatic poles, which may give rise to tufted dendrites (*small arrows*), and a small cell (*large arrow*). **D** Small neurons

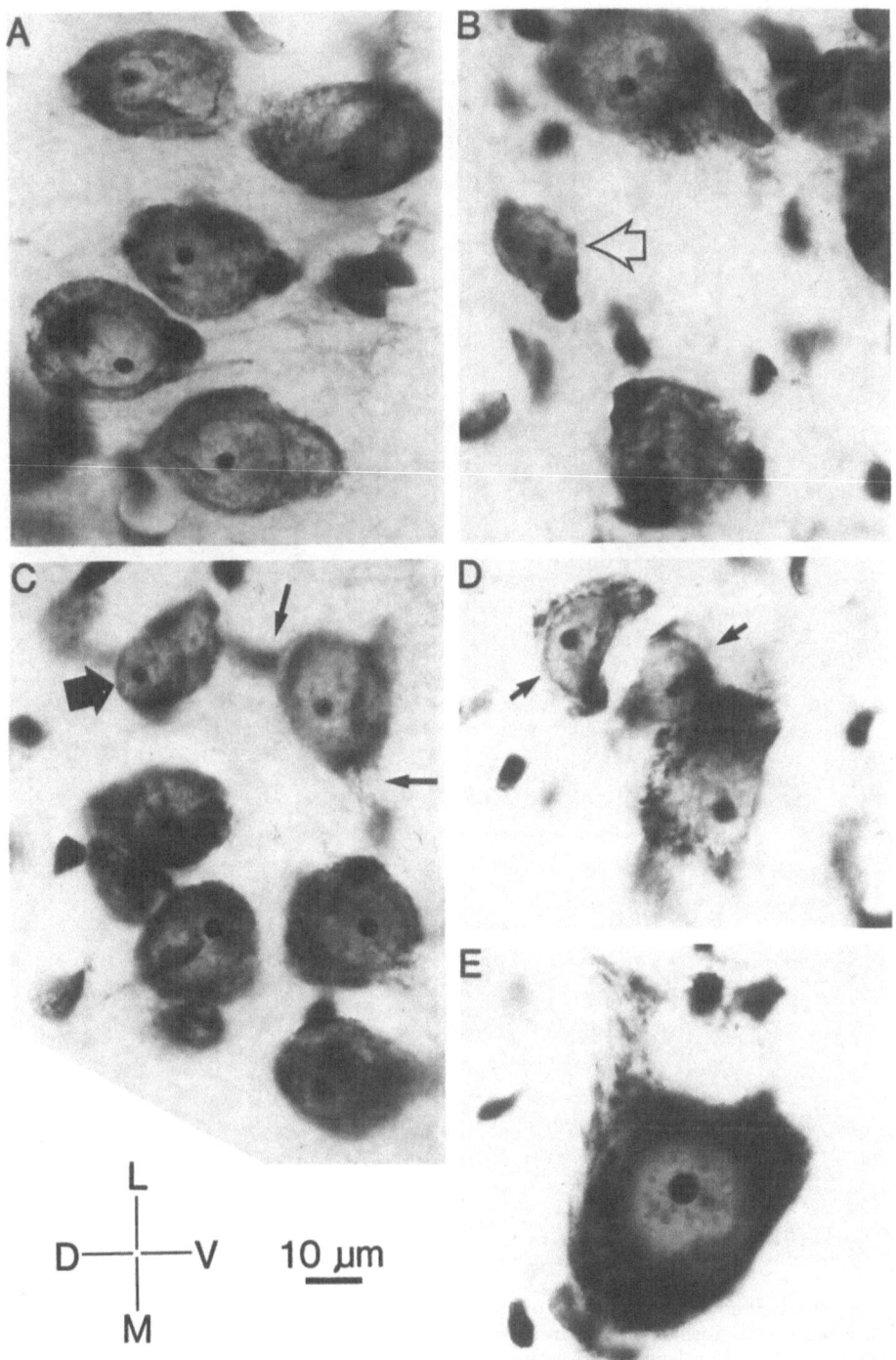

(*arrows*) and a medium-sized cell from the medial division; the dendrites of the latter are polarized along the medial-to-lateral axis. **E** Magnocellular neuron from the medial division. Adult cat, 15-μm-thick, paraffin-embedded transverse section; no adjustment was made for shrinkage in this or Fig. 45. Semi-apochromat, N.A. 1.25, ×787.5

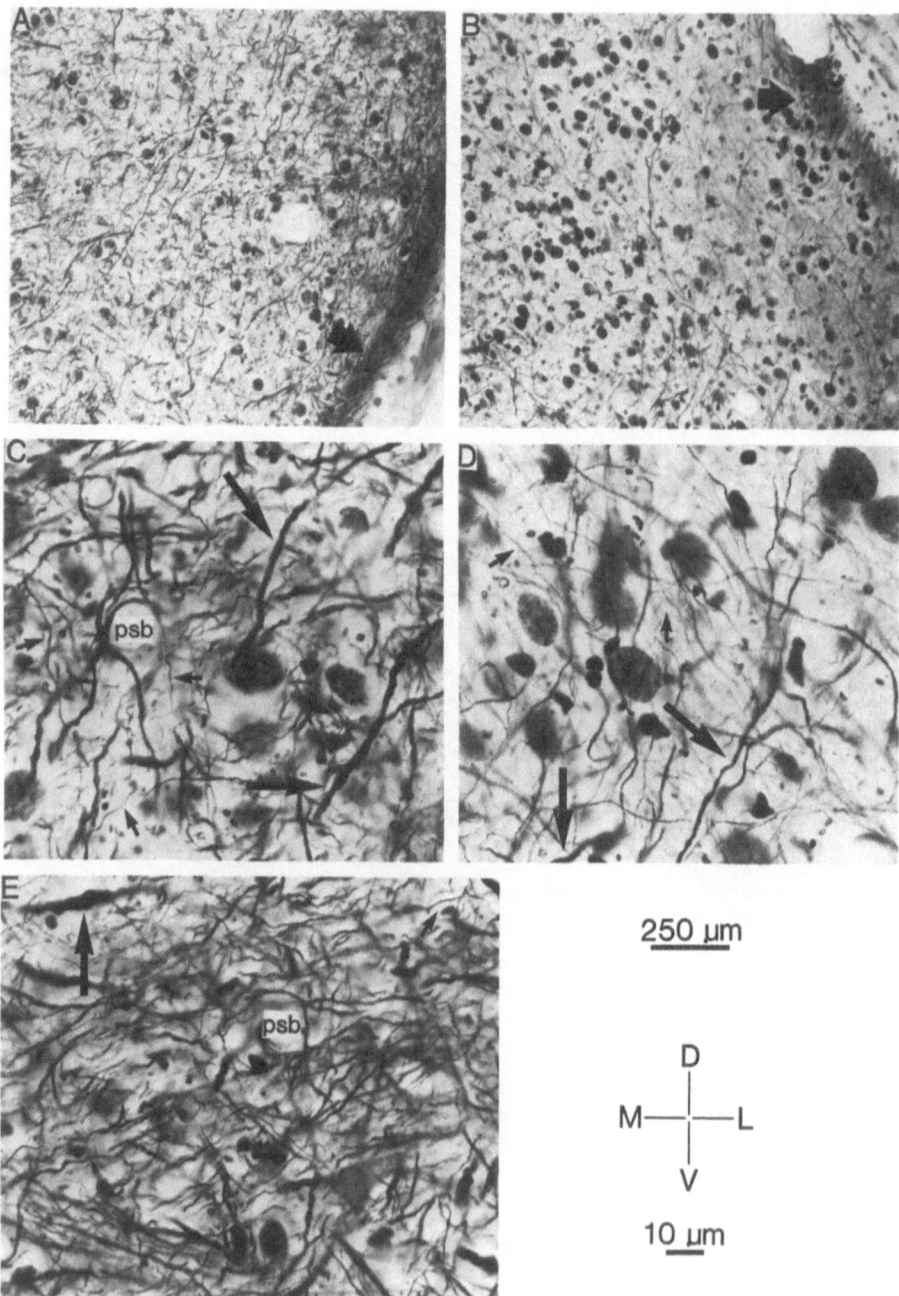

250 μm

D
M——|——L
V

10 μm

Fig. 45 A–E. Photomicrographs of transversely sectioned Bodian preparations from medial geniculate body nuclei. **A** Low-power view of the lateral edge of the ventral division, just dorsal to the ventrolateral nucleus (see Fig. 1 A) and including the marginal zone (*arrow*). Note the ventral-to-dorsal arc of ascending axons. **B** Low-power view of the fiber architecture of the superficial dorsal and dorsal nuclei, which has a fine, lacy appearance. *Arrow,* marginal zone. For **A, B:** adult cat, 15-μm-thick, paraffin-embedded transverse section; planapochromat, N.A. 0.14, ×50. Upper scale bar. **C** High-power view of axons in *pars lateralis*, showing coarse (*larger arrows*) and fine (*small arrows*) axons, some forming perisomatic baskets (*psb*).

68

above. The spines are variable and even encroach upon the proximal portions of the shafts. The axon resembles that of the cells described above.

The magnocellular neuron (Fig. 39) is the largest, and least common, cell in the medial division. It occurs primarily in more rostral sections (Fig. 2C, D) where it is conspicuous. It has a large (35 × 40 μm), round soma and five-to-eight major dendrites. Often the dendritic field is the same size or smaller than that of wide-field (Fig. 36) or medium-sized radiate (Fig. 38) cells. The arbors of this cell, while large, are irregularly shaped, and the dendrites branch only once or twice before tapering to fine endings as they fill a spherical field. Dendritic appendages predominate on the central shafts and intercept a variety of endings (Fig. 39: *1, 2*). The distal dendrites, like those of wide-field (Fig. 36), medium-sized bushy (Fig. 37: *stippled outline*), and medium-sized stellate (Fig. 38) cells, often overlap or wind about each other. Emerging from the soma, the thick axon projects only a few micrometers and acquires a myelin sheath as it curves toward the acoustic radiations. No axon collaterals have been seen to arise from this axon.

The only intrinsic axon in the medial division with a locally projecting collateral system is from *the Golgi type II cell* (Fig. 40). This neuron is scattered throughout the medial division and is the smallest cell there; it is among the smallest in the entire medial geniculate body with its 10 × 15 μm, round or triangular soma. The four-to-five slender, main dendrites branch at least twice, often parallel to each other and the plexus of brachial axons, and form a fine, dense arbor of terminal segments. The appendages of these sparsely spinous cells are thin, sometimes pedunculated, and primarily limited to the central parts of the shafts. Often the dendrites are no thicker than the axon of this cell. The axon originates from the soma and within 30 μm emits several collaterals which branch acutely; many thin segments pass from the plane of section, and others taper and end as fine (less than 0.5-μm-thick), twiglike terminals (Fig. 40, upper left).

Associated with the medial division, particularly along its caudal part, where it is not fully developed, are the neurons of the interstitial nuclei of the brachium of the inferior colliculus (Fig. 41). These elongated cells have sparsely branched dendrites and are scattered throughout, and conform to the orientation of, brachial fibers. The interstitial nuclei (Figs. 2A, 3A) form the caudomedial border of the dorsal nucleus. The expansion of the medial division (Fig. 2B, C) defines the rostral limit of the interstitial nuclei. Cells in the interstitial nuclei have two or three main dendrites issuing from the somatic poles. The long axis of the dendrites is crossed by brachial axons (Fig. 41: *1, 2*) which enter from the ventromedial edge of the medial geniculate body. The disposition of the axon of the cell is unknown (see also Winer and Morest 1983b).

Many kinds of brachial and extrabrachial axons enter the medial division. It is difficult to determine if these axons project only to the medial division,

◁ **D** High-power view of medium-sized (*larger arrows*) and extremely thin (*small arrows*) axons in the superficial dorsal nucleus; perisomatic baskets are absent. **E** High-power view of medial division axons near the brachium of the inferior colliculus to show very thick (*larger arrow*) and thinner (*small arrow*) axons and occasional perisomatic baskets (*psb*). For **C, D, E**: adult cat, 15-μm-thick, paraffin-embedded section; semi-apochromat, N.A. 1.25, × 787.5. Lower scale bar give off collaterals while passing toward more lateral and dorsal regions, or traverse the medial division and make few, if any, synaptic contacts. Since the brachial axons have a tortuous course, it is unusual to see their complete preterminal and terminal fields in continuity. Extrinsic axons in the medial division include medium-sized, beaded fibers (Fig. 36: *1*); thicker, perisomatic terminals observed exclusively on magnocellular neurons (Fig. 39: *1*); a variety of very thin axons, some with beaded swellings or other specializations; and other, smooth axons (Fig. 42 B). Besides the large perisomatic, calyxlike ending, which has two or three separate branches and superficially resembles the type IV ending in the dorsal nuclei (Fig. 23), the medial division contains a distinctive axon not otherwise seen in the medial geniculate body. These complex, serpentine terminals occupy and dominate substantial volumes of neuropil in the medial division and have a very thick (3–4 µm) central branch which emits up to 50 collaterals which ramify in constricted spirals near the origin (Fig. 42 A: *1, 2*). These axons form dense, nestlike structures covering the distal dendrites of principal cells. The caliber and the complexity of their collateral system distinguishes them from Golgi type II axons, but their provenance is unknown (Fig. 40).

The neuropil of the medial division has a distinctive texture dominated by medio-laterally oriented brachial axons (Figs. 43, 45 E). Scattered among these axons are magnocellular neurons (Fig. 43: *stippled outline* above *M*) and small fascicles of dorso-ventrally projecting axons; the latter may be of cortical origin, since their size and course are like presumptive corticothalamic axons in the dorsal division (Fig. 25). The arrangement of axons is so distinctive in the medial division that it can be differentiated, on the basis of its fiber architecture, from the adjacent deep dorsal nucleus (Fig. 43: *DD*).

The ascending projections of the medial division include every subdivision of auditory cortex, without obvious topography. Even the smallest injections of horseradish peroxidase in a cortical area (Fig. 13: *primary auditory cortex*) retrogradely label a few cells in the medial division, and larger injections label more neurons (Winer et al. 1977). The pattern of labeled neurons also differs in the medial division. Thus the labeled cells are scattered throughout large parts of the medial division after injections of every subdivision of auditory cortex (Figs. 13, 31–34). The cortical target of the medial division extends beyond the auditory areas (Kuroda et al. 1975; Spreafico et al. 1981). It is likely that several of the different types of medial division neurons, including some of the smallest cells, project to auditory cortex. After large horseradish peroxidase injections which mark nearly all the neurons in the ventral nucleus, surprisingly large numbers of medial division cells are also labeled. In the most heavily labeled cases large, medium-sized, and small somata are routinely filled with peroxidase granules. Thus, at least three of the five types of cells in the medial division – the magnocellular neurons, one or more varieties of the medium-sized cells, and certain of the small cells possibly corresponding to the Golgi type II cell – project to auditory cortex (Winer 1984b).

4 Discussion

4.1 Anatomical Correlates of Physiological Function

The physiological implications of *ventral division* anatomy were described in some detail by Morest (1975a, b) and are briefly reviewed here as a basis for considering the other subdivisions of the medial geniculate body. The orderly registration of frequency was demonstrated behaviorally by Ades et al. (1939). Later physiological work showed that neurons in the medial geniculate body were sensitive to acoustic stimuli (Gross and Thurlow 1951). The presence of a tonotopic map was confirmed by Rose and Woolsey (1958), who stimulated different parts of the cochlea and found that the apex is represented laterally in the ventral division, and the base medially. The structure of these tonotopic maps conforms to the shape and disposition of ventral division laminae (Aitkin and Webster 1972; de Ribaupierre and Toros 1974). These electrophysiological mapping studies showed that unit tuning curves ranged from broad to narrow and that units responsive to low frequency had wider tuning curves than high-frequency cells.

The morphologic basis for at least part of this map is undoubtedly the regular arrangement of principal neurons and the laminar afferent axonal plexus in the ventral division. Ascending axons, probably from the central nucleus of the inferior colliculus (Table 2; Kudo and Niimi 1980), project to the central, spine-covered portions of both principal and Golgi type II cells (Jones and Rockel 1971). They may project also, though presumably less strongly, to the principal neuron soma. The intermediate dendrites receive sparse synaptic endings from auditory corticofugal axons. Most input is to the distal dendrites of principal thalamic neurons (Morest 1975a). A more complex, local arrangement is elaborated by the axons of Golgi type II cells, which project to principal cell dendrites, and whose dendrites form dendrodendritic junctions with principal cell dendrites (Morest 1971, 1975a). These dendrites are the locus for the synaptic nest consisting of presynaptic Golgi type II dendrites, principal cell dendrites, and colliculogeniculate and Golgi type II axons, all surrounded by a glial sheath (Morest 1971, 1975a). This may be a common feature of thalamic nuclei (Ralston 1971).

That the bulk of the dendritic tree of single neurons is confined to a single lamina is consistent with the physiological data showing sharply tuned best frequencies for most units in the ventral nucleus (Aitkin and Webster 1972). Since the axonal plexus within the ventral nucleus contributes to this laminar organization (Fig. 7:2), a morphological basis exists for precise tuning over most of the unit's response area and for less sharply tuned, but still tonotopically organized, flanks beside the narrowly tuned center.

The various combinations of cortical and midbrain input, along with differences in the intrinsic organization of the ventral division, somehow underlie

Table 2. Summary of connections of the medial geniculate body[a]

	Reference
I. Ascending connections from the midbrain and caudal brain stem	
A. Projections from the inferior colliculus	
1. Projection from central nucleus to ventral division	Jones and Rockel (1971); Oliver and Hall (1978a)[b]
2. Projection of pericentral nucleus (posterior cortex) to dorsal division	Kudo and Niimi (1980)
B. Projections from lateral tegmental nuclei of the midbrain	
1. Tegmentogeniculate and inferior parabrachial inputs to the dorsal nucleus	Morest (1965b); Oliver and Hall (1978a)[b]; Winer and Morest (1984)
2. Tegmentogeniculate and (2 varieties of) inferior parabrachial inputs to the deep dorsal nucleus	Morest (1965b); Winer and Morest (1984)
3. 4 separate, overlapping axonal inputs from lateral tegmental sources to suprageniculate nucleus	Morest (1965b); Winer and Morest (1984)
C. Projection of central acoustic tract	
1. From the vicinity of the superior olivary complex or nuclei of the lateral lemniscus to the medial division	Papez (1929a); Jones (1979)[b]; Henkel (1983)
D. Other ascending inputs to the medial geniculate body	
1. Vestibular nuclear input to medial division	Mickle and Ades (1954); Raymond et al. (1974); Roucoux-Hanus and Boisacq-Schepens (1977)
2. Spinothalamic tract projection to medial division	Mehler et al. (1960)[c]; Berkley (1980)
3. Superior colliculus projection to medial division	Graham (1977)
II. Thalamofugal connections	
A. Cortical projections of the ventral division	
1. Projection of ventral and ovoid nuclei to primary auditory cortex	Rose and Woolsey (1949); Winer et al. (1977)
2. Projection of ventral nucleus to anterior auditory field	Andersen et al. (1980b)
B. Cortical projections of the dorsal division	
1. Projections of superficial dorsal and ventrolateral nuclei to the posterior ectosylvian fields and the second auditory cortex	Winer et al. (1977)
2. Projection of deep dorsal nucleus to the second auditory cortex	Winer et al. (1977)
3. Projection of the caudal dorsal nucleus to temporal cortex	Winer et al. (1977)
4. Projections of the suprageniculate nucleus to insular cortex	Winer et al. (1977)

Table 2 [continued]

	Reference
C. Projections of the medial division	
1. Projections to all subdivisions of auditory cortex	Winer et al. (1977)
2. Projections beyond auditory cortical areas	Kuroda et al. (1975); Spreafico et al. (1981)
D. Subcortical projections	
1. From *pars lateralis* to putamen	Ebner (1967[d], 1969[d]); Ryugo and Killackey (1974)[e]
III. Corticothalamic projections	
A. Projections to ventral division	
1. Projections from primary auditory cortex	Diamond et al. (1969)
2. Projections from non-primary auditory cortex	Pontes et al. (1975)
B. Projections to dorsal division	
1. Projections from auditory cortex to dorsal superficial, dorsal, and deep dorsal nuclei	Diamond et al. (1969)
2. Projections from non-auditory cortex	Graybiel and Berson (1980)
C. Projections to medial division	
1. Projections from auditory cortex	Pontes et al. (1975)
2. Projections from non-auditory cortex	Jones and Powell (1973)

[a]   Unless otherwise indicated, observations are based on the cat
[b]   Tree shrew (*Tupaia glis*)
[c]   Rhesus monkey (*Macaca mulatta*)
[d]   Opossum (*Didelphis virginiana*)
[e]   Rat

the complex and diverse temporal response patterns in single units generated by acoustic stimuli. Thus units are excited, or inhibited, or have long, reverberatory responses to tones (Aitkin et al. 1966). The complex, waxing and waning responses may be intrinsic to the medial geniculate body, since stimulation of either the brachium of the inferior colliculus or the auditory cortex evoked reverberatory responses, and ablation of the cortex failed to abolish them (Aitkin and Dunlop 1969; see also Kallert and Kroha 1978). In Morest's (1975a) model, the excitatory colliculogeniculate endings converge on synaptic nests and are driven by afferent volleys. This excitation also activates the (presumably inhibitory) Golgi type II cell, whose axon and dendrites are presynaptic to principal cells. Since the axon collaterals of Golgi type II cells are relatively numerous, an excitatory volley (or series) will eventually result in an interneuron-mediated inhibition after an initial excitatory response. Because the Golgi type II axons may project to different synaptic nests than their dendrites (Morest 1975a), the influence of these cells might be propagated to comparatively large areas of the medial geniculate body. The first arrangement has been studied in the electron microscope (as the synaptic nest, or glomerulus) by Morest (1971, 1975b). The second and more diffuse arrangement extends the influence of Golgi type II neurons, albeit in attenuated form, to the dendrites of principal cells remote from the synaptic nests and may influence these distant nests in a sparsely distributed fashion rather than the concentrated fashion characteristic of the immediate vicinity of the local circuit neuron. Inhibitory and some purely excitatory interactions have been confirmed by intracellular recording (Nelson and Erulkar 1963). Powerful corticothalamic projections (Andersen et al. 1980b; see Table 2) whose physiological consequences are unknown are another mechanism influencing medial geniculate neurons (Ryugo and Weinberger 1976). The influence that any of these different pathways have on the shape of more complex physiological events within the medial geniculate body is unknown, e.g., corticogeniculate endings contain pleomorphic vesicles (Morest 1975a) whose synaptic effects are not known.

Binaural responses of medial geniculate body cells have been known for many years (Galambos 1952; Galambos et al. 1952). An excitatory-inhibitory interplay was shown to be sensitive to binaural stimulation (Adrián et al. 1966). Most units were activated by stimulation of the contralateral ear, some by both ears, and the fewest by ipsilateral stimulation. Simultaneous binaural stimulation may produce either larger (facilitation) or smaller (occlusion) effects than the monaural response. Unequal binaural activation may elicit inhibitory responses otherwise masked by symmetric binaural input (Aitkin and Dunlop 1969; Aitkin and Webster 1972). Among the mechanisms that might influence these binaural effects are midbrain (or cortical) inputs to particular nuclei and the intrinsic organization of the nucleus. Since the input from the central nucleus of the inferior colliculus to the ventral division of the medial geniculate body is bilateral and extensive (Moore and Goldberg 1963; Andersen et al. 1980a), it probably contributes to these response properties. Perhaps excitatory-excitatory binaural interactions primarily reflect inferior colliculus inputs, while excitatory-inhibitory sequences embody these and also reflect the (presumptively inhibitory) influences of local circuit neurons as well as corticothalamic projections. At present, however, this scenario is speculative.

In the *dorsal division* at least one, and perhaps more, tonotopic maps exist. Electrode penetrations crossing the ventral division latero-medially thus show an orderly relationship between tonotopic progressions of best frequency and electrode position. There is a dramatic difference at the border of the dorsal division such that best frequency responses occur in loose clusters, with considerable scatter, and little evidence of a systematic tonotopic trend. As the electrode appears to enter the suprageniculate nucleus there is a shift in frequency (generally toward higher frequencies), and, as scatter increases, units often form clusters of similar frequency spanning intervals of several kHz (see Woolsey 1972, citing unpublished observations of Gross et al. [1963]; see also Morel and Imig 1982). From the sparse available data for the dorsal division, it cannot be judged how many tonotopic representations of frequency exist or if such map(s) each comprise the complete frequency range of the basilar membrane. Even if several maps exist, their organization differs fundamentally from the frequency map in the ventral division.

The diversity of axonal input to the dorsal division could imply a corresponding complexity of electrophysiological responses. Thus, in the cat these neurons respond to auditory and somatic inputs (Lippe and Weinberger 1973a, b) and in the rabbit also to visual stimuli (unpublished observations of Stewart et al., cited in Lippe and Weinberger 1973a, b). Many different physiological response profiles are seen in the dorsal nuclei. Dorsal division cells have low rates of spontaneous activity and high sensitivity to ambient sound. Neurons with prolonged reverberatory responses are common in the dorsal nucleus (Aitkin et al. 1966). Response areas with interleaved, excitatory-inhibitory profiles occur in both unanesthetized (Aitkin and Prain 1974) and freely moving animals (Whitfield and Purser 1972). Other patterns include cells with irregular, bursting spike trains, or broad tuning curves, and cells which habituate to repeated tones (Aitkin 1973; Aitkin and Prain 1974).

Substrates exist which could account for some of these physiological findings. Thus the tufted principal cell dendrites and their extensive innervation by axonal types I and II might subserve whatever tonotopic organization exists in the dorsal nuclei. The size and peridendritic distribution of type IV axons could also provide a basis for such a map in view of their size and rather close proximity to the axon hillock on principal cells. The broader tuning curves could belong to the equally abundant principal stellate neurons, whose dendrites might receive afferents from a much larger sector of the neuropil. Perhaps these neurons embody a parallel, less tonotopically organized pathway in the dorsal division. The complex, prolonged excitatory-inhibitory interactions and the relatively low rates of spontaneous activity might reflect the influence of the two classes of Golgi II cells and the auditory cortical input (Orman and Humphrey 1981). The numerous intrinsic cells and their many collaterals provide a background for local influence on afferent activity, but the exact morphological basis for these actions remains largely speculative.

The three main sources of extrinsic input to the dorsal division are the inferior colliculus, the lateral tegmental system of the midbrain, and the cerebral cortex (Table 2). These functionally diverse areas are probably the origin, respectively, of much ascending auditory (Kudo and Niimi 1980) and somatic sensory (Morest 1965b) input to the dorsal nuclei. The cortical input to these nuclei is possibly inhibitory (Ryugo and Weinberger 1978) and originates from widespread areas (Diamond et al. 1969). Since at least two types of neurons in the dorsal nuclei have locally arborizing axons, the eight classes of extrinsic axons provide many opportunities for contact, both direct and indirect, with principal cell dendrites (and perhaps with Golgi type II dendrites too). In electron micrographs, many axodendritic and axospinous, and few axosomatic, synaptic endings occur on principal cells (Winer and Morest 1984, Fig. 15; Winer unpublished observations). If the influence of these interneurons (and/or the cerebral cortex) is inhibitory, it could sustain the prolonged time course of waxing and waning or spindling behavior of neurons in the dorsal nuclei, events suggestive of local, reverberatory neuronal circuits. The convergence of the different kinds of afferent axons on the same principal cells is consistent with this sequence of long, complex physiological events, as is the diverse structure and intermingling of these types of neurons. If specializations like the synaptic nests or glomeruli of the ventral division (Morest 1971, 1975a, b) exist also in the dorsal division, then comparable fine structural arrangements may be present, although the diversity of the intrinsic axonal plexus suggests that other synaptic patterns might occur. At present it is not certain if the corticogeniculate projection has excitatory or inhibitory consequences in the medial geniculate body or some combination of these (Ryugo and Weinberger 1978; Orman and Humphrey 1981). In any event, it appears to have more diverse effects than the visual cortex-lateral geniculate body pathway, which is believed to have excitatory synaptic effects (Kalil and Chase 1970).

In the *medial division* the heterogeneity of the nerve cells and the diversity of axons passing through (and terminating) suggest that the acoustic functions in other parts of the medial geniculate body must here undergo a radical transformation (Winer 1979; Winer and Morest 1983b). No clear cytoarchitectonic subdivision can be made within the medial division, for its structurally diverse cells lie scattered in a complex neuropil through which brachial and extrabrachial axons pass. The functional properties of medial division neurons differ significantly from dorsal and ventral division cells. While a weak tonotopic organization is present in the medial division, these units have broader tuning curves and are less selective for acoustic stimuli than cells in other subdivisions (Aitkin 1973; Lippe and Weinberger 1973a, b). These neurons respond to input from several modalities, including the vestibular, visual, and somatic sensory systems (Wepsic 1966). In addition, the thalamic reticular nucleus appears to project to the medial and dorsal (but not to the ventral) divisions of the medial geniculate body, which could admit non-auditory influences (Jones 1975).

4.2 Comparison with Other Thalamic Nuclei

The neurons of the medial geniculate body are similar to neurons in many thalamic sensory nuclei. While certain cells may be morphologically specialized and rare in other thalamic nuclei, e.g., the strongly tufted bushy neuron of the ventral division or the large Golgi type II cell in the dorsal nuclei, many neurons in the medial geniculate body are like those in other thalamic subdivisions. Thus bushy principal cells of the dorsal nuclei resemble neurons in the anterior nuclei of the feline thalamus (Somogyi et al. 1979) and in the rodent (McAllister and Wells 1981) and cat (Ramón y Cajal 1911) ventrobasal complex, and they are like pyramidal cells of the hippocampus in certain respects (Lorente de Nó 1934).

The stellate principal neurons of the dorsal and suprageniculate nuclei are among the most common cell types in the thalamus. Comparable neurons occur in the cat ventrobasal complex (Scheibel and Scheibel 1966a) and lateral geniculate body (O'Leary 1940; Guillery 1966), in the primate pulvinar (Ogren and Hendrickson 1979), and in the midline thalamic nuclei of carnivores (Leontovich 1975) and primates (Hazlett et al. 1976). The geometry of their dendrites suggests that these neurons, compared with bushy cells, have a different functional role in processing auditory information.

Neurons like the small Golgi type II cell in the ventral and dorsal divisions are also found throughout the thalamus: in the ventrobasal complex (Scheibel et al. 1972), in the lateral geniculate body (Famiglietti and Peters 1972), in the pulvinar (Ogren and Hendrickson 1979), in certain non-specific thalamic nuclei (Scheibel and Scheibel 1966a, 1967), and in various species. Neurons resembling the large Golgi type II cell occur in the lateral geniculate body (Friedlander et al. 1981). The principal neuron of the posterior limitans nucleus is like certain cells in the reticular formation (Scheibel and Scheibel 1958; Ramon-Moliner and Nauta 1966), in the thalamic reticular nucleus (Scheibel and Scheibel 1966b), and in various brain stem and other centers in the cat (Ramon-Moliner 1969). It is not unreasonable to suggest that these neurons, or the other classes of cells described above, might share certain common functional features; indeed, their shapes may be of heuristic value in predicting function.

4.3 Comparative Anatomy of the Medial Geniculate Body

Little is known about the structure of the medial geniculate body (or its homologue) in non-mammalian species. A more complete review than the following brief account is available in Morest and Winer (to be published). In both reptiles (Pritz 1974; Browner and Rubinson 1977) and birds (Karten 1967; Leibler 1976) a large midbrain auditory nucleus projects to the thalamus, which in turn projects to a telencephalic acoustic center which may be homologous to the auditory cortex (Leppelsack 1974; Saini and Leppelsack 1977, 1981).

The structure of nuclei presumed to be homologous to the medial geniculate body has been examined primarily in Nissl or fiber preparations (Papez 1929b, 1936; Haight and Neylon 1978). In mammals there have been few investigations of the structure of the medial geniculate body in species other than the cat. A study by Oliver (1982; see Table 1) in the tree shrew identified many of the nuclei described above and revealed certain interspecific differences, e.g., the tree shrew medial division can be divided into a rostral and a caudal part on the bases of structure and connections, and the laminae of the ventral nucleus have a somewhat different orientation than in the cat. In spite of these variations and other differences, the basic plan embodied by these species appears relatively similar and provides a basis for comparative studies of fine structure with reference to specialized arrangements, for example, the synaptic glomeruli or nests in the ventral nucleus. The arrangement of neurons in certain subdivisions, e.g., bushy principal and Golgi type II cells in the ventral nucleus, suggests some basic congruences between tree shrew and cat which are preserved also in the pattern of thalamocortical connections (Oliver and Hall 1978b).

A comparative study of the cat and opossum revealed many similarities in neuronal architecture (Morest 1965c; Winer and Morest 1979; Morest and Winer to be published). Thus the ventral division in both species contains cells with bushy dendrites and a tufted branching pattern; these cells are arranged in parallel rows and project to the cerebral cortex (Winer et al. 1977; Winer and Coleman unpublished observations). A small Golgi type II cell is also present in both species. Comparable cell types with similar dendritic arrangements occur in the dorsal and medial divisions. Despite these similarities, there are significant interspecific differences. In the cat dorsal nuclei, as many as 40% of the cells have extensive, locally branching axons forming an extensive plexus in the neuropil (Winer and Morest 1984). In the opossum, such Golgi type II cells are much less common, the contribution of their axon collaterals to the neuropil is commensurately reduced, and the synaptic organization of the medial geniculate body might be fundamentally different. There is also a qualitative distinction in the degree of dendritic specialization. Thus, bushy principal neurons in the feline ventral nucleus are highly tufted, while the dendrites of the corresponding cell in the opossum are much less so. This or analogous patterns are consistent throughout the nuclei of the medial geniculate body. In primates, the laminar distribution of thalamocortical afferents from the medial geniculate body and associated nuclei appears comparable to the pattern in the cat, as do the main cytoarchitectonic features, although the latter point awaits elaboration with morphological techniques (Burton and Jones 1976; Jones and Burton 1976).

The major morphological features of medial geniculate body organization in the cat are found in each species studied, although certain differences are apparent. Thus the laminar structure of the ventral division is conspicuous, and enormously expanded, in both species of bats studied (*Antrozous pallidus* and *Pteronotus p. parnellii*) and less obvious in the hedgehog (*Erinaceus europaeaus*). The fibrodendritic laminae are remarkably clear in the raccoon (*Procyon lotor*) and somewhat less so in the monkey (*Macaca mulatta*) due, in part, to the elaboration of the intralaminar neuropil. Stellate or bushy principal cells and small neurons are present in the various nuclei of the medial geniculate body of these species (Winer unpublished observations). The neuronal architecture of the human medial geniculate body is readily comparable to the cat. Despite the fact that in humans the laminae in the ventral nucleus assume a much more medio-lateral orientation than in the cat, the principal bushy neurons in both are quite similar in their dendritic configuration and comparative size, and smaller stellate neurons with radiate branching patterns are also evident. Similar neurons occur also in the nuclei of the dorsal division in humans, including bushy and stellate cells with large dendritic fields and smaller stellate neurons which may be related to intrinsic pathways. These similarities do not imply identical principles of organization or comparable neural circuits, for it is clear by other criteria, e.g., the elaboration of the human neuropil and the reduction in neuronal packing density, that the similarities in dendritic configuration or cytoarchitecture in no way confirm functional equivalence or imply homology (Winer 1984c). The latter depends on complex, interrelated criteria, including neural structure, position, developmental history, synaptic architecture, and common ancestral origin — dimensions not yet thoroughly assessed in any species. The phylogeny of the medial geniculate body in most non-mammalian species has been studied only in cell- and fiber-stained material (Clark 1933; Papez 1936). Thus definitive conclusions about homologous neural populations based on dendritic and axonal structure and patterns of connections cannot yet be drawn (Morest 1965c; Winer and Morest 1979; Morest and Winer to be published).

4.4 Functional Organization of the Medial Geniculate Body

The main point of this paper is that the medial geniculate body cannot be regarded as one nucleus with a single function. It is a mosaic of nuclei which differ radically in structure and connections and probably represent many functions. The specialized anatomical and physiological organization of the ventral division suggest that it may be primarily auditory compared with the other subdivisions. But describing a nucleus as "auditory" does not define its function. While a rigid tonotopic map of frequency is found in the ventral division, this does not imply that frequency analysis is the only or primary function, or that it has a single function. Besides responding to frequency information, the ventral division must have some role in analyzing binaural input and in spatial localization of sound, in encoding of loudness and timbre information, and perhaps in higher integrative function (Erulkar 1975; Aitkin 1976). In each sensory modality, representation at the thalamic (and cortical) level is now thought to include multiple, adjoining maps (Merzenich and Kaas 1980). These representations and re-representations of the various sensory epithelia must serve other functions than merely to repeat the pattern of discharge encoded by receptors and the central propagation of these impulses. Some speculations on the different patterns of integration for each subdivision of the auditory thalamus are given below. Many of these ideas are present or implicit in modified form or are anticipated in other papers (e.g., Moore and Goldberg 1963; Graybiel 1972a, b, 1973; Casseday et al. 1976; Oliver and Hall 1978a, b; Andersen et al. 1980b; Graybiel and Berson 1981).

The structure, connections, and physiology of the ventral division define it as the target and chief component of the primary (lemniscal) auditory pathway. Within the ventral division there is a proportional representation of frequency consistent with the tonotopic organization of the basilar membrane and recapitulating it, albeit in expanded form. This expansion may be due partly to the elaboration of interneurons and their axonal processes. The excitatory or inhibitory interplay of these cells could serve a delaying and filtering function and may be involved in the temporal analysis of sound (Rouiller et al. 1979). The physiological data in the primary auditory cortex are consistent with the idea of little or no convergence in the projection of thalamic neurons (Phillips and Irvine 1981). Considering the size of the auditory input to ventral division cells, it would be surprising to find an equally powerful, non-auditory input to them. It is unknown if there is any map of acoustic space in the ventral division.

A principal target of the cortical projection of the ventral division is the primary auditory cortex, where binaural columnar bands are present through most of the middle- and high-frequency representation (Imig and Adrián 1977; Middlebrooks et al. 1980; see also Imig and Reale 1981). It is unknown if comparable columns exist in the medial geniculate body or if the convergence of thalamic (Sousa-Pinto 1973) and commissural and/or corticocortical input (Imig and Reale 1980, 1981) creates such cortical columns. Since few systematic recordings have been made parallel to isofrequency contours or within single laminae of the ventral division, this question remains unanswered (see de Ribaupierre and Toros 1974). That there may be different patterns of binaural selectiveness in the medial geniculate body is suggested by physiological experiments in which contralateral dominant or suppression areas are found in the ventral division (Calford and Webster 1981).

The dorsal division is predominantly, but not exclusively, auditory in function. Its several nuclei may contain multiple representations of frequency (Woolsey 1972; Morel and Imig 1982). Its cytologically diverse structure and widespread cortical projections could compare inputs from auditory and other sensory systems, and perhaps coordinate auditory-visual and auditory-somatic or other reflexes. The large network of intrinsic Golgi type II axons and the collateral branches of afferent axons are mechanisms for polymodal convergence on single cells. It is not known which dimensions of acoustic and other sensory information are compared, or how the comparison is effected. Since some spatial segregation of tectogeniculate and corticogeniculate input occurs on principal cell dendrites in the ventral division (Jones and Rockel 1971; Morest 1975a), a comparable situation might exist in the dorsal division, particularly on principal cells with bushy dendrites. The extensive innervation by type I axons to the central dendritic segments of both stellate and bushy principal cells could convey some tonotopic influence in these nuclei, as noted above. In fact one might predict that bushy neurons are more likely to have narrow, and stellate cells somewhat broader, tuning curves, a suggestion consistent with the correspondence noted between dendritic structure and function in other neural systems (e.g., Kolb and Normann 1982). The many different midbrain inputs to, and cortical targets of, dorsal division cells reinforce the idea that its nuclei function pluralistically and that the separate but interrelated sensory channels remain segregated at the cortex (Winer et al. 1977; Reale and Imig 1980; Graybiel and Berson 1981).

The strongest evidence for this segregation is the selective nature of the thalamocortical inputs from different dorsal division nuclei and the tonotopic organization of each of the cortical territories receiving dorsal division fibers. Thus at least seven auditory cortical fields (Fig. 5) besides the primary auditory cortex have a tonotopic representation (Woolsey 1961; Reale and Imig 1980). Each field receives a somewhat different pattern of thalamocortical input from the dorsal division (Winer et al. 1977; Niimi and Matsuoka 1979). For example, the ventrolateral nucleus projects exclusively to the posterior ectosylvian gyrus, the deep dorsal nucleus sends most of its axons to the second auditory cortex, and the caudal extremity of the dorsal nucleus projects chiefly to the cortex behind the pseudosylvian sulcus. More examples of this specificity could be adduced (see Table 2). This does not imply that there is no overlap in the cortical projections of the different dorsal division nuclei or that the target of a single neuron is necessarily limited to a single cortical tonotopic-architectonic-functional field. Rather, it suggests that the tonotopy (or binaural responsiveness) in the cortex is unlikely to result only from ipsilateral corticocortical fibers (Winguth and Winer 1983) or commissural fibers (Code and Winer 1983) and that it must depend to some degree on the specificity and spatial segregation of inputs from particular thalamic nuclei, or parts of them.

The tonotopic organization of frequency in dorsal division nuclei may impose some specificity in non-primary auditory cortical areas (Imig and Reale 1981), although these maps may, in part, reflect patterns of corticocortical connections (Kawamura 1973; Paula-Barbosa et al. 1975). That the sharpness of tuning and frequency representation in non-primary (or peri-) auditory cortex may be less precise and orderly than in primary cortex (Hind 1953; Merzenich et al. 1975; Irvine and Huebner 1979; Reale and Imig 1980) suggests that the tonotopic map(s) in the dorsal division may also be less regular than in the ventral division, or that the overlap in the cortical projections of the separate dorsal division nuclei collectively obscures what is otherwise an orderly map. In view of the relatively broader tuning curves of dorsal division neurons (Aitkin and Prain 1974), the tonotopic map(s) in the dorsal nuclei might be expected to be less exact, regular, or complete than those in the ventral division; nonetheless, the possibility of collateral overlap at the cortical level cannot be excluded without double labeling or similar experiments (see Irvine and Huebner 1979). Insofar as the dorsal nuclei may be considered to project to primary auditory cortex, their functional role is uncertain. Numerically, their input to primary auditory cortex is a fraction (less than 7% of the total number of medial geniculate body cells projecting to physiologically identified EE [excitatory-excitatory] or EI [excitatory-inhibitory] binaural bands) of that from the ventral nucleus, whose cells provide more than 90% of the thalamic input to primary auditory cortex. Thus, the impact of the dorsal division on the maintenance of, for example, a tonotopic map in primary auditory cortex is uncertain (Middlebrooks and Zook 1983).

The dorsal division of the present study includes numerous thalamic territories previously designated under the collective rubric of the "posterior group" (Rose and Woolsey 1958; Jones and Powell 1971; Graybiel 1972b). For the most part, these and other prior accounts of the posterior group have subdivided the posterior thalamus into posterior lateral, intermediate, and medial (or comparable) nuclei essentially on the basis of different patterns of connections, secondarily on the basis of cytoarchitecture, and with little regard for the subtle but nonetheless consistent differences in neuronal architecture which, together with the connectional and architectonic data, form the core of the present account. An example of a correlation between morphological and connectional methods is provided by the ventrolateral nucleus, which is here regarded as belonging to the dorsal division on these grounds, in spite of the fact that it lies completely within the confines of the ventral division. Its large, multipolar stellate neurons occupy a small wedge of tissue between the mass of the ventral nucleus and the marginal zone, and these cells project exclusively to non-primary auditory cortex, areas dominated by input from various other dorsal division nuclei (Winer et al. 1977). This connectional heterogeneity appears to be a common principle of dorsal division organization, where spatially separate thalamic territories may project convergently onto individual cortical fields. It may well be that these cells differ also in their electrophysiological properties.

It is proposed that the ventral division consists of the ventral and ovoid nuclei and the marginal zone − each with a common neuronal architecture and each projecting only to primary auditory cortex (and the anterior auditory field). The limits of the term "dorsal division" must also be enlarged to include the posterior limitans nucleus, whose cells have previously not been separately considered or may have been subsumed as part of the posterior group or "Po". The form of these cells is heterotypical within the medial geniculate body and in general they resemble the weakly stellate neurons usually found in the thalamic reticular nucleus or the brain stem (Ramon-Moliner and Nauta 1966; Ramon-Moliner 1969) rather than the other, more characteristic bushy or stellate dendritic configuration of the thalamic neurons described here. But they, too, project to the auditory cortex and are labeled after massive injections of horseradish peroxidase saturating the primary and adjoining auditory cortex (Winer, unpublished observations). A second reason for including them in the dorsal division is their physical proximity to, and many *boutons de passage* from, midbrain, lateral tegmental, and corticofugal axons destined to terminate in other, more lateral parts of the medial geniculate body. The striking morphology of these cells and, indeed, the morphologically distinct nature of each of the nuclei described in this paper argue that the diverse cell types and patterns

81

of thalamocortical connections are related to each other in a functional way. Subdividing the thalamus primarily on geographic, connectional, or cytoarchitectonic bases would tend to minimize many of the distinctions forming the essence of the present account. It is tempting to speculate that much the same connectional and architectonic approach led to the formulation of the concept of primary sensory and association cortex which has now proliferated into a veritable banquet of cortical subdivisions (Merzenich and Kaas 1980; Woolsey 1981). There is no *a priori* reason to suppose that the number of subdivisions of thalamic (and midbrain or hindbrain) sensory nuclei is much greater or much smaller than the number of cortical subdivisions until we know more about each of these structures. But in many areas different in their physiology and connections, e.g., areas 17 and 18, it is not surprising also to find a corresponding set of basic morphological differences (Lund 1973; Lund et al. 1981). It is therefore likely that even the many subdivisions of the medial geniculate body described here may be susceptible to further functional, synaptic, or histochemical partition (e.g., Graybiel and Berson 1980; Middlebrooks and Zook 1983).

In the medial division, auditory input, while imposing some tonotopic organization, is not dominant, and the maps of frequency are less orderly (and have been less systematically explored) than elsewhere in the medial geniculate body. The disposition of medial division neurons with respect to the brachial (and extrabrachial) fibers suggests that its chief function may not be point-to-point fidelity in the transmission of frequency information. As the only part of the medial geniculate body which projects to all subdivisions of auditory cortex (Winer et al. 1977) and to adjoining non-auditory cortical areas (Kuroda et al. 1975), it is likely to have some different function other than the maintenance of tonotopicity. Since medial division input to the cerebral cortex has a different laminar target than ventral division axons (layers I and VI versus layers IIIb–IV [Sousa-Pinto 1973; Niimi and Naito 1974]), the information conveyed by medial division neurons remains segregated at least until the first cortical synapse. The medial division also provides some input to the amygdala, thus linking auditory and/or non-auditory pathways with the subcortical telencephalon (Russchen 1982).

As a preliminary and provisional attempt to stimulate a more developed idea of medial geniculate body functions, the following proposal is offered (see Table 3). Ventral division cells are likely to be active in the fine-grained, discriminative aspects of auditory perception. It is predicted that the limited number of cell types probably receive a relatively homogeneous afferent input and that the synaptic endings are rigidly segregated in serial order and establish a tonotopic sequence upon the dendrites of the principal cells, which systematically compare the input from related but different sources along, for example, laterally versus medially placed dendrites. These dendrites may receive input from somewhat differentially tuned axons. Extrinsic circuits here may exert considerable influence. Feedback or feedforward neural loops might be arranged in parallel to compare afferent input across adjoining, as well as within, laminae. The laminae are suited also to the spatial and temporal analysis of sound, which would be supported by the lattice-like, periodical arrangement of principal and Golgi type II cells and their stereotyped synaptic relations. It is hypothesized that lesions of the ventral division alone (if such were possible) would disturb

Table 3. Summary and functional comparison of medial geniculate body divisions

Division	Proposed function	Number of sub-divisions	Number of cell types	Variety and size of afferents	Sharp-ness of afferent tuning	Inter-neu-ronal plexus	Possible effect of lesion
Ventral	Acoustic analysis	3	2	Moderate; medium-sized and thick	High	Moder-ately developed	Loss of specific discriminative capacity
Dorsal	Acoustic attention	10[a]	8	More; thin and medium-sized	Varied	Highly developed	Sensory inattention
Medial	Multi-sensory arousal	1	5	Most; thin-to-thick	Broad	Poorly developed	Little or no modality-speci-fic sensory loss

[a] Includes anterior dorsal nuclei

discriminative or analytical appreciation, e. g., speech perception or the detection of signals embedded in noise — in short, the *relations among* complex trains of acoustic stimuli.

Neurons in the dorsal division, on the other hand, may have a broader role in acoustic behavior, perhaps related to attention in the auditory thalamus, to the regulation or coordination of cyclic activity from various sensory systems, but among which acoustic input nonetheless is dominant. The wide range in the morphological specialization of its cells implies a correspondingly broad spectrum of function: the bushy principal cells most closely resemble ventral nucleus bushy neurons and may have analogous functional responses, e. g., narrow tuning curves and a strong preference for auditory input. Other, more "generalized" cells, e. g., the stellate neurons, might be expected to have broader tuning curves and polysensory responses; diverse, probably overlapping synaptic input to their dendrites; and perhaps extrinsic afferents with disparate physiological effects as well. No large-scale laminar organization comparable to the ventral nucleus is expected, and the magnification factor of the tonotopic map(s) may bear little, if any, relation to auditory behavioral acuity. The regular terminal pattern of axonal contacts common in the ventral nucleus and presumed to propagate the tonotopic map of frequency is reduced here and convergent; multisensory input to principal cell dendrites is common. The extreme development of the interneuronal axon plexus could underlie rapid, diffuse conduction to select neural elements, exciting some and simultaneously inhibiting others, and activating several (but not necessarily all) modalities. It is hypothesized that damage to the dorsal division alone might leave discriminative acoustic function(s) intact but disrupt the selective action and direction of auditory attention, perhaps leading to an auditory inattention. Here it is proposed that the *relations between* different sensory pathways among different neurons would be perturbed.

In the medial division, the plurality of cell types, the heterogeneous afferent axonal plexus, and lack of laminar dendritic or axonal arrangements suggest that there is likely to be little, if any, spatial segregation of auditory input comparable to ventral nucleus fibrodendritic laminae. Particular cells might receive convergent or multimodal input, among which auditory axons are only one of many. Broad sensory tuning curves and the modest intrinsic axonal plexus could serve the diffuse, prolonged, non-specific "avalanche-type" propagation of impulses to broad territories in the auditory thalamus and beyond. Axonal endings here rarely form the regular, sequential patterns noted in the ventral nucleus. The functional implication is that damage confined to the medial division would reduce the efficacy of polysensory arousal, and a corollary is that it would be the most difficult lesion of the three described here to detect since its pathways and influence through the auditory and extraauditory thalamus are so diffuse. Perhaps the efficiency or integration of acoustic-motor reflexes would be subtly impaired. It is thus likely that the *relations across* sensory pathways are compromised by such damage. These speculations are intended to be provocative, reductionistic, and heuristic and as prolegomena to neurobehavioral research.

There is no unified or single theory of medial geniculate body function because it is a complex of nuclei with diverse structures and several functions. In this sense it resembles the trigeminal system — which is not a nucleus but a complex of nuclei (Darian-Smith 1973). There is no *a priori* reason to expect that a particular physiological function (e.g., frequency analysis) is itself unitary or is represented in, or performed in the same way, by all nuclei or each subdivision. The structural and physiological heterogeneity of the medial geniculate complex suggests that some functions may indeed be segregated and others shared. The presence of parallel sensory pathways (e.g., multiple tonotopic representations) does not imply that each of these is equivalent; how they are alike and how unique remains to be determined. The potential segregation of function in the nuclei of the medial geniculate body is not unique to the auditory thalamus. For example, the ventral division is part of a pathway whose origins probably extend at least to the cochlear nucleus. This pathway is represented and propagated in the midbrain by the disc-shaped neurons and fibrodendritic laminae in the central nucleus of the inferior colliculus. These neurons are prominent in the central nucleus of each species examined in this study (Winer 1981, and in preparation). Their laminar arrangement might coincide with an orderly arrangement of frequency (or other dimensions) in this nucleus. The fidelity of this pathway is preserved in the ventral nucleus of the medial geniculate body, which receives projections from the central nucleus of the inferior colliculus (Andersen et al. 1980a), and where comparable maps of frequency are found (Aitkin and Webster 1972; Morel and Imig 1982). These narrowly tuned auditory neurons project, in turn, to a limited portion of the auditory cortex, AI (and to the anterior auditory field [Andersen et al. 1980b]). They may end near the tufted and stellate cortical neurons whose columnar organization (Rose 1949) and polarized dendritic arbors (Winer 1982, 1984a) are reminiscent of the disc-shaped cells of the central nucleus of the inferior colliculus and of the bushy neurons of the ventral division of the medial geniculate body. These columns are by no means to be considered as the final, common sensory pathway in the auditory cortex, for they undoubtedly contribute to the further elaboration of this channel along commissural, corticocortical, and corticofugal pathways influencing other acoustic nuclei and, indirectly, themselves (Winer 1984a; see also Jones and Powell 1970).

Some recent work suggests that the ventral nucleus may embody more than one parallel pathway to the primary auditory cortex. Thus Middlebrooks and Zook (1983) found that excitatory-excitatory or excitatory-inhibitory cortical bands receive discontinuous input from different ventral nucleus sectors, implying that these physiologically distinct cortical areas might result from functionally distinct pathways in the ventral nucleus. An additional pathway might include the projection from certain small cells in the ventral nucleus to AI (Winer 1984b). The ventral nucleus might therefore have at least four inputs to AI: some bushy neurons projecting to one kind of binaural band, others to a different kind, while still others (in lower-frequency parts of AI) have a projection pattern different from the first two (cf. Middlebrooks and Zook 1983, Fig. 11); finally, the small stellate cell projection completes the picture. There is no obvious structural difference within the ventral nucleus, except the coiled fibrodendritic laminae in the ovoid nucleus.

Besides the electrophysiological, morphological, and connectional heterogeneity of the medial geniculate body, the subdivisions can be further differentiated on the basis of their affinity for acetylcholinesterase (Graybiel and Berson 1980). Thus parts of the suprageniculate-lateral posterior complex stain intensely for this enzyme and receive input from auditory (Diamond et al. 1969) and periauditory cortical areas (Graybiel and Berson 1980, 1981). This contrasts with the ventral division, which receives a major input from primary auditory cortex and stains only palely with this enzyme. However, choline acetyltransferase-positive neurons or terminal fields occur in the ventral nucleus and parts of the dorsal and medial divisions (Kimura et al. 1981), those in the ventral nucleus being most intensely stained. Medial geniculate neurons are also reported to respond to iontophoretically applied acetylcholine (Tebēcis 1972). This suggests that suprageniculate-lateral posterior neurons may share histochemical, as well as morphological, features with each other and associated dorsal division nuclei (see also Jones and Powell 1971; Graybiel 1972a, b, 1973; Winer and Morest 1983b).

The primary auditory pathway includes portions of the cochlear nucleus, the central nucleus of the inferior colliculus, the ventral nucleus of the medial geniculate body, and the primary auditory cortex. It coincides in large measure with the lemniscal pathway defined by Graybiel (1972b, 1973) on connectional grounds. The integrity of this pathway is probably essential for complex acoustic discriminations (Jerger et al. 1969). The pathways embodied in the dorsal and medial divisions are not exclusively auditory. It is probable that the functional role of some of their neurons has little to do with highly discriminative aspects of hearing, particularly since animals with lesions in these areas recover rather quickly in behavioral auditory discrimination tasks (Glassman et al. 1975). Most of these cells, however, respond to auditory stimuli, project to auditory cortex, and receive inputs from these non-primary acoustic areas of the brain stem and cerebral cortex. They are thus auditory without being exclusively acoustic. These secondary pathways bear a certain resemblance to Graybiel's concept of the lemniscal adjunct system (Graybiel 1972b, 1973), but they differ in ways enumerated below. A premise of the present paper is that the morphologically and connectionally distinct nuclei of the dorsal division are likely to have individual roles in auditory and non-auditory function. The idea that they may be ancillary or secondary to the so-called primary pathway should not imply that their organization is any less developed than that of the primary pathway or that their projection pattern or any segregation of synaptic terminals is any less exact or that they function collectively in some obscure, amalgamated way.

There is a long tradition in neuroanatomy of classifying thalamic sensory nuclei and their cortical projection fields by cytoarchitecture and neuronal organization, connections, and functional disposition (see Winer and Morest, 1983 b, for a more detailed account of the history). One root of this quest is the effort to explain the evolution and organization of cortical areas with a heavy thalamic input (e.g., primary sensory cortex) and cortical areas apparently devoid of, or receiving only a sparse number of, thalamic fibers (e.g., non-primary or association cortex). It was postulated that these areas receive, respectively, essential or sustaining thalamic projections on the bases of their different tonotopic organization and susceptibility to retrograde degeneration (Rose and Woolsey 1949, 1958). Later, Diamond and Chow (1962) and Diamond and Hall (1969) proposed that a core of primary sensory cortex was ringed by a belt of non-primary cortex and that both types of cortex receive thalamic input. Graybiel (1972b, 1973) elaborated the functional and connectional concepts of the lemniscal line and the lemniscal adjunct systems. The lemniscal line pathway was conceived as recently evolved, as having high synaptic security and fidelity to afferent input, and as extending from the sensory periphery to the primary sensory neocortex with limited synaptic interruption and highly topographic organization. The lemniscal adjunct system was believed to be older, to have lower synaptic security and less fidelity to afferent input, and to have a limited topographic organization, polymodal afferent and efferent connections, and perhaps more synaptic stations too.

Andersen et al. (1980b) also described two ascending pathways to the subdivisions of the auditory cortex. Their cochleotopic system projects to primary auditory cortex, the anterior auditory field, and various sectors of the posterior ectosylvian gyrus (see Fig. 5B, C). The target of the diffuse system includes the second auditory cortex and the cortex behind the pseudosylvian sulcus. They concluded that both systems (a) contain more than one functionally defined auditory area, (b) that the systems of (thalamocortical and corticothalamic) connections correspond to two functionally different types of auditory fields, and (c) that the topographic patterns of the two systems are different, the diffuse system having other rules of organization than the cochleotopic. These conclusions raise a number of questions relating both to the thalamus and to the cerebral cortex. First, is the medial division, which projects to all subdivisions of auditory cortex and beyond, part of the diffuse system or a separate system in its own right, or are there degrees of diffuseness? Second, does the fact that single units have broad tuning curves imply that the projections to them are necessarily diffuse (i.e., of diverse origin) or that the locus of individual synaptic inputs to the cell is randomly distributed? Third, do the eponyms "cochleotopic" and "diffuse" imply a continuum or a dichotomous distribution of functional properties? For example, many of the AI cells innervated by axons arising from either the dorsal or the medial divisions probably have broad tuning curves, and in this sense the cochleotopic and diffuse systems are superimposed in the same cortical area, although they remain segregated on a laminar basis. Fourth, in what neurobehavioral sense do the cochleotopic or lemniscal line (or segments of them) and the diffuse or lemniscal adjunct systems have different or complementary roles in auditory discriminations? In-

86

sofar as they have separate functional roles, damage to them could have differential effects. Fifth, how is a cortical field to be defined? The anterior auditory field, for example, immediately adjoins the primary auditory cortex but has a reversed tonotopic sequence of best frequencies (Knight 1977; Andersen 1979; Andersen et al. 1980b). Despite this difference, it receives thalamocortical input from what are apparently the same tonotopically organized sectors of the ventral nucleus which project topographically upon the primary auditory cortex, it projects reciprocally onto the same thalamic subdivisions from which ascending fibers to it originate, and the tuning curves of individual units are reported to be indistinguishable (except for the reversed tonotopy) from those in AI. Perhaps the AAF-AI boundary is somehow analogous to the horizontal meridian delimiting the upper and lower visual fields: a geographic border of an otherwise continuous sensory map. Sixth, are tonotopic organization and patterns of connections sufficient criteria for subdividing cortical (or thalamic) areas? For example, somata in various dorsal division nuclei retrogradely labeled from horseradish peroxidase injections in the non-primary auditory cortex are widely, but not randomly, distributed, and there is a certain specificity in the pattern of projection which, while apparently less orderly than the ventral nucleus-AI pathway, is not nearly as unpredictable as the projections of the medial division. Multiple tonotopic maps in the dorsal nuclei — a possibility consistent with its neuronal diversity and heterogeneous afferents (Aitkin et al. 1981) — cannot be discounted (see Morel and Imig 1982). That the pattern of corticothalamic projections from non-primary auditory cortex appears to obey the principle of reciprocity and specificity in the AI-ventral nucleus circuit is indirect evidence against the diffuse or adjunct character of these nuclei. The architecture and functional organization of dorsal division neurons, though different from the ventral or medial divisions, is probably no less specific or specialized. Indeed, the argument has been advanced that in certain respects these neurons and their cortical projection fields may have evolved *after* the so-called cochleotopic or lemniscal system and that their broader tonotopy and polymodal responsiveness is characteristic of phylogenetically advanced neurons (Morest and Winer to be published).

A second argument for reconsidering the evolutionary status of the dorsal division (and the cortex it innervates) is that its cells are structurally comparable in many ways to cells in the subdivisions of the pulvinar (Winer and Morest 1983b). While its nuclei have for many years been considered as phylogenetically and functionally advanced (Walker 1938), they receive projections from brain stem nuclei (e.g., the superior colliculus) with a much older history. Since the pulvinar projects chiefly to non-primary visual cortex (Harting et al. 1972; Glendenning et al. 1975), it also links older and newer limbs of a sensory pathway. Perhaps the subdivisions of the inferior colliculus projecting to the dorsal division of the medial geniculate body are older than those projecting to the ventral nucleus (Andersen et al. 1980a; Aitkin et al. 1981). Neurons in certain portions of the dorsal division (notably suprageniculate cells) and in the medial division are similar in structure to many cells in the pulvinar-lateral posterior complex (Hajdu et al. 1974). Certain of the latter neurons have been characterized as the unaligned cells of the posterior thalamus, whose functional properties are also thought to differ fundamentally from those of the main (or lemniscal) sensory nuclei of the thalamus (see Winer and Morest 1983b, Fig. 12).

These secondary channels can be contrasted with the spinothalamic input to the thalamus. While representations of the body surface occur in the posterior thalamus (Berkley 1980), they are large, frequently bilateral, and less topographically organized than those in the ventrobasal complex (Poggio and Mountcastle 1960, 1963). Analogous plans have been described for the retinogeniculate and retinotectal visual systems (Schneider 1969; Graybiel 1972b, 1973; Tusa et al. 1978; Graybiel and Berson 1980). The medial division must be considered, on the basis of its connections and the morphology of its neurons, as part of the medial geniculate body. It comprises a plurimodal, parallel pathway beside the primary auditory pathway of the ventral nucleus, and the polymodal pathways in the dorsal division of the medial geniculate body, together in the auditory cortex, where they are further interrelated by corticocortical and commissural connections. The influence of the medial division on the cortex is thus likely to be broad and only one of many; it is singular in projecting so widely. As with the dorsal division, however, this hypothesis does not suggest imprecision or an indiscriminate blending of these axonal endings. Rather, it implies a pattern which is distinct from those of the ventral and dorsal divisions, both in terms of the structure of its cells (Winer and Morest 1983b) and their differential laminar terminations (Niimi and Naito 1974). Further levels of complexity could be superimposed as, for example, in the massive, widely distributed serotonergic input to the rat supragranular frontoparietal cortex (Descarries et al. 1975). The origin and termination of other fine, broadly projecting afferents to the auditory cortex (and perhaps to the medial geniculate body) is unknown (see Morrison et al. 1978; Itakura et al. 1981). These pathways, compared with those of the ventral, dorsal, and medial divisions, may indeed be diffuse.

The assignment of more precise and differential functions to the nuclei of the medial geniculate body awaits physiological and behavioral studies. At present the primary pathway in the ventral division is one extreme, while the medial division appears to be an auditory analogue of the brain stem reticular formation in several respects (Ramon-Moliner and Nauta 1966) or may even be related to the intralaminar nuclei of the thalamus. The dorsal division is analogous to the retinotectal-pulvinar visual pathway in its breadth of inputs, diversity of cell types, polymodal (though predominantly acoustic) function, and plurality of cortical targets. The nuclei of the medial geniculate body can be subdivided into a primary auditory pathway (the ventral division), a system of polymodal associated nuclei (the dorsal division), and a plurimodal pathway (the medial division). These parallel and partly overlapping channels in the auditory thalamus recapitulate, in miniature, the organization of the thalamus.

5 Summary

The neuronal organization of the cat medial geniculate body was studied to harmonize its cytoarchitecture, neuronal architecture, the pattern of cortical projections, and its functional organization. The primary finding is that the medial geniculate body is a mosaic of nuclei assignable to three divisions which differ in form, connections, and probably function. A second conclusion is that each division can be further subdivided into several nuclei: in the case of the ventral division, into three nuclei; in the dorsal division, into five nuclei; the medial division is one nucleus with regional architectonic differences. These nuclei each contain characteristic types of cells and differ in their brain stem and cortical connections and may have separate tonotopic maps. The ventral division contains two types of cells: principal neurons with bushy dendrites and a smaller cell with a stellate dendritic configuration and a locally arborizing axon. In the dorsal division, principal cells with stellate, bushy, or fusiform dendrites are common, and four types of small cells with intrinsic axons have been observed. In the medial division, four varieties of principal cells and a single kind of interneuron occur. Differences in the organization and texture of the axonal plexus in the neuropil and the individual course, size, and branching patterns of afferent and intrinsic axons further distinguish these divisions and their nuclei. The differential pattern of thalamocortical projections from (and midbrain and cortical inputs to) each subdivision extends their independent but interrelated organization. Thus the ventral division receives ascending auditory input mainly from the central nucleus of the inferior colliculus and projects chiefly to primary auditory cortex. The dorsal division has diverse auditory and non-auditory inputs from various midbrain nuclei and projects broadly upon several areas of non-primary auditory cortex. The medial division receives heterogeneous sensory input from several midbrain and brain stem nuclei and projects widely to all subdivisions of auditory cortex and to adjoining cortical areas. No single or simple theory can account for medial geniculate body function, nor is a single function attributable to any nucleus. It is proposed that the three divisions of the auditory thalamus represent a specific sensory lemniscal pathway (the ventral division), a polymodal non-primary association pathway (the dorsal division), and a plurimodal pathway (the medial division).

6 Acknowledgements

These studies were supported by USPHS Grants 1 NSO5432 and 1 RO1 NS16832, and by University of California Faculty Research Grants. It is a pleasure to acknowledge the technical assistance of L. Andrus, C. Bellue, S. D'Amato Flynn, H. Do, T. Simon, and R. Williams. I am grateful to J.C. Middlebrooks and D.L. Oliver for their comments on an earlier version of the manuscript.

7 References

Ades HW, Mettler FA, Culler EA (1939) Effect of lesions in the medial geniculate bodies upon hearing in the cat. Am J Physiol 125:15–23

Adrián HO, Lifschitz WM, Tavitas RJ, Galli FP (1966) Activity of neural units in medial geniculate body of cat and rabbit. J Neurophysiol 29:1046–1060

Aitkin LM (1973) Medial geniculate body of the cat: responses to tonal stimuli of neurons in medial division. J Neurophysiol 36:275–283

Aitkin LM (1976) Tonotopic organization at higher levels of the auditory pathway. In: Porter R (ed) Neurophysiology II (International review of physiology vol 10) University Park Press, Baltimore, pp 249–279

Aitkin LM, Dunlop CW (1969) Inhibition in the medial geniculate body of the cat. Exp Brain Res 7:68–83

Aitkin LM, Prain SM (1974) Medial geniculate body: unit responses in the awake cat. J Neurophysiol 37:512–521

Aitkin LM, Webster WR (1972) Medial geniculate body of the cat: organization and responses to tonal stimuli of neurons in ventral division. J Neurophysiol 35:365–380

Aitkin LM, Dunlop CW, Webster WR (1966) Click-evoked response patterns of single units in the medial geniculate body of the cat. J Neurophysiol 29:109–123

Aitkin LM, Calford MB, Kenyon CE, Webster WR (1981) Some facets of the organization of the principal division of the cat medial geniculate body. In: Syka J, Aitkin L (eds) Neuronal mechanisms of hearing. Plenum Press, New York, pp 163–181

Andersen RA (1979) Patterns of connectivity of the auditory forebrain of the cat. Thesis, University of California

Andersen RA, Roth GL, Aitkin LM, Merzenich MM (1980a) The efferent projections of the central nucleus and the pericentral nucleus of the inferior colliculus in the cat. J Comp Neurol 194:649–662

Andersen RA, Knight PL, Merzenich MM (1980b) The thalamocortical and corticothalamic connections of AI, AII, and the anterior auditory field (AAF) in the cat: evidence for two largely segregated systems of connections. J Comp Neurol 194:663–701

Ariëns Kappers CU, Huber GC, Crosby EC (1936) The comparative anatomy of the nervous system of the vertebrates, including man. Macmillan, London

Berkley KJ (1980) Spatial relationships between the terminations of somatic sensory and motor pathways in the rostral brain stem of cats and monkeys. I. Ascending somatic sensory inputs to lateral diencephalon. J Comp Neurol 193:283–317

Bodian D (1937) The staining of paraffin sections of nervous tissue with activated Protargol. The role of fixatives. Anat Rec 69:153–162

Bourk TR, Mielcarz JP, Norris BE (1981) Tonotopic organization of the anteroventral cochlear nucleus of the cat. Hear Res 4:215–241

Browner RH, Rubinson K (1977) The cytoarchitecture of the torus semicircularis in the tegu lizard, *Tupinambis nigropunctatus*. J Comp Neurol 176:539–558

Burton H, Jones EG (1976) The posterior thalamic region and its cortical projection in new world and old world monkeys. J Comp Neurol 168:249–302

Calford MB, Webster WR (1981) Auditory representation within principal division of cat medial geniculate body: an electrophysiological study. J Neurophysiol 45:1013–1028

Cant NB, Morest DK (1979) The bushy cells in the anteroventral cochlear nucleus of the cat. A study with the electron microscope. Neuroscience 4:1925–1945

Casseday JH, Diamond IT, Harting JK (1976) Auditory pathways to the cortex in *Tupaia glis*. J Comp Neurol 166:303–340

Clark WE, Le Gros (1933) The medial geniculate body and the nuclei isthmi. J Anat 67:536–548

Code RA, Winer JA (1983) Heterogeneous origin of commissural neurons in cat primary auditory cortex (AI). Proc Soc Neurosci 9:953

Cowan WM, Gottlieb DI, Hendrickson A, Price JL, Woolsey TA (1972) The autoradiographic demonstration of axonal connections in the central nervous system. Brain Res 37:21–51

Cox W (1891) Impregnation des centralen Nervensystems mit Quecksilbersalzen. Arch Mikr Anat 37:16–21

Darian-Smith I (1973) The trigeminal system. In: Iggo A (ed) Somatosensory system. Handbook of sensory physiology vol II. Springer, Berlin Heidelberg New York, pp 271–314

de Ribaupierre F, Toros A (1974) Analyse unitaire de l'activité neuronale du corps genouille médian en réponse à une stimulation acoustique. Bull Acad Sci Méd 30:118–123

Descarries L, Beaudet A, Watkins KC (1975) Serotonin nerve terminals in adult rat neocortex. Brain Res 100:563–588

Diamond IT, Chow KL (1962) Biological psychology. In: Koch S (ed) Psychology: a study of a science. vol 4, Study II. McGraw-Hill, New York, pp 158–241

Diamond IT, Hall WC (1969) Evolution of neocortex. Science 164:251–262

Diamond IT, Jones EG, Powell TPS (1969) The projection of the auditory cortex upon the diencephalon and brain stem in the cat. Brain Res 15:305–340

Ebner FF (1967) Medial geniculate projections to telencephalon of opossum. Anat Rec 157:238–239

Ebner FF (1969) A comparison of primitive forebrain organization in metatherian and eutherian mammals. Ann NY Acad Sci 167:241–257

Erulkar SD (1975) Physiological studies of the inferior colliculus and medial geniculate complex. In: Keidel WD, Neff WD (eds) Auditory system. Handbook of sensory physiology vol V, part 2. Springer, Berlin Heidelberg New York, pp 145–198

Famiglietti EV Jr, Peters A (1972) The synaptic glomerulus and the intrinsic neuron in the dorsal lateral geniculate nucleus of the cat. J Comp Neurol 144:285–344

Friedlander MJ, Lin C-S, Stanford LR, Sherman SM (1981) Morphology of functionally identified neurons in lateral geniculate nucleus of the cat. J Neurophysiol 46:80–129

Galambos R (1952) Microelectrode studies on medial geniculate body of cat. III. Responses to pure tones. J Neurophysiol 15:381–400

Galambos R, Rose JE, Bromiley RB, Hughes JR (1952) Microelectrode studies on medial geniculate body of cat. II. Responses to clicks. J Neurophysiol 15:359–380

Glassman RB, Forgus MW, Goodman JE, Glassman HN (1975) Somesthetic effects of damage to cats' ventrobasal complex, medial lemniscus or posterior group. Exp Neurol 48:460–492

Glendenning KK, Hall JA, Diamond IT, Hall WC (1975) The pulvinar nucleus of *Galago senegalensis*. J Comp Neurol 161:419–458

Golgi C (1878) Un nuovo processo di tecnica microscopia. RC Inst Lombardo Sci 2nd series 12:5

Golgi C (1879) Di una reasione apparentemente nera della cellule nervose cerebrali ottenuta col bicloruro di mercurio. Arch Sci Med (Torino) 3:1–7

Golgi C (1891) Modificazione del metodo di colorazione deli elementi nervosi col bicloruro di mercurio. Riv Med Napoles 7:193–194

Golgi C (1906) The neuron doctrine-theory and facts. In: Nobel lectures: Physiology or medicine, 1901–1921. Elsevier, Amsterdam (1967 reprint) pp 189–217

Graham J (1977) An autoradiographic study of the efferent connections of the superior colliculus in the cat. J Comp Neurol 173:629–654

Graybiel AM (1972a) Some ascending connections of the pulvinar and nucleus lateralis posterior of the thalamus in the cat. Brain Res 44:99–125

Graybiel AM (1972b) Some fiber pathways related to the posterior thalamic region in the cat. Brain Behav Evol 6:363–393

Graybiel AM (1973) The thalamocortical projection of the so-called posterior nuclear group: a study with anterograde degeneration methods in the cat. Brain Res 49:229–244

Graybiel AM, Berson DM (1980) Histochemical identification and afferent connections of subdivisions in the LP-pulvinar complex and related nuclei in the cat. Neuroscience 5:1175–1238

Graybiel AM, Berson DM (1981) On the relation between transthalamic and transcortical pathways in the visual system. In: Schmitt FO, Worden FG, Adelman G, Dennis SG (eds) The organization of the cerebral cortex. MIT Press, Cambridge, pp 285–319

Gross NB, Thurlow WR (1951) Microelectrode study of neuronal auditory activity of cat. II. Medial geniculate body. J Neurophysiol 14:409–422

Guillery RW (1966) A study of Golgi preparations from the dorsal lateral geniculate nucleus of the adult cat. J Comp Neurol 128:21–50

Haight JR, Neylon L (1978) An atlas of the dorsal thalamus of the marsupial brush-tailed possum, *Trichosurus vulpecula*. J Anat 126:225–245

Hajdu F, Somogyi G, Tömböl T (1974) Neuronal and synaptic arrangement in the lateralis posterior-pulvinar complex of the thalamus in the cat. Brain Res 73:89–104

Harting JK, Hall WC, Diamond IT (1972) Evolution of the pulvinar. Brain Behav Evol 6:424–452

Hazlett JR, Dutta CR, Fox CA (1976) The neurons in the centromedian-parafascicular complex of the monkey (*Macaca mulatta*): a Golgi study. J Comp Neurol 168:41–74

Henkel CK (1983) Evidence of sub-collicular auditory projections to the medial geniculate nucleus in the cat: an autoradiographic and horseradish peroxidase study. Brain Res 259:21–30

Hind JE (1953) An electrophysiological determination of tonotopic organization in auditory cortex of cat. J Neurophysiol 16:475–489

Imig TJ, Adrián HO (1977) Binaural columns in the primary field (A1) of cat auditory cortex. Brain Res 138:241–257

Imig TJ, Reale RA (1980) Patterns of cortico-cortical connections related to tonotopic maps in cat auditory cortex. J Comp Neurol 192:293–332

Imig TJ, Reale RA (1981) Ipsilateral corticocortical projections related to binaural columns in cat primary auditory cortex. J Comp Neurol 203:1–14

Irvine DRF, Huebner H (1979) Acoustic response characteristics of neurons in nonspecific areas of cat cerebral cortex. J Neurophysiol 24:107–122

Itakura T, Kasamatsu T, Pettigrew JD (1981) Norepinephrine-containing terminals in kitten visual cortex: laminar distribution and ultrastructure. Neuroscience 6:159–175

Jerger J, Weikers NJ, Sharbrough FW III, Jerger S (1969) Bilateral lesions of the temporal lobe. Acta Otolaryngol [Suppl] (Stockh) 258:1–51

Jones DR (1979) Auditory pathways in the brain stem of the tree shrew, *Tupaia glis*. Thesis, Duke University

Jones EG (1975) Some aspects of the organization of the thalamic reticular complex. J Comp Neurol 162:285–308

Jones EG, Burton H (1976) Areal differences in the laminar distribution of thalamic afferents in cortical fields of the insular, parietal, and temporal regions of primates. J Comp Neurol 168:197–248

Jones EG, Powell TPS (1970) An anatomical study of converging sensory pathways within the cerebral cortex of the monkey. Brain 93:793–820

Jones EG, Powell TPS (1971) An analysis of the posterior group of thalamic nuclei on the basis of its afferent connections. J Comp Neurol 143:185–216

Jones EG, Powell TPS (1973) Anatomical organization of the somatosensory cortex. In: Iggo A (ed) Somatosensory system. Handbook of sensory physiology vol II. Springer, Berlin Heidelberg New York, pp 579–620

Jones EG, Rockel AJ (1971) The synaptic organization in the medial geniculate body of afferent fibers from the inferior colliculus. Z Zellforsch 113:44–66

Kalil RE, Chase R (1970) Corticofugal influence on activity of lateral geniculate neurons in the cat. J Neurophysiol 33:459–474

Kallert S, Kroha E (1978) Reversible Kaltblockade der primären Hörrinde bei der wachen Katze. Naturwissenschaften 65:211–212

Karten HJ (1967) Organization of the ascending auditory pathway in the pigeon (*Columba livia*). I. Diencephalic projections of the inferior colliculus (nucleus mesencephalicus lateralis, pars dorsalis). Brain Res 6:409–427

Kawamura K (1973) Cortico-cortical fiber connections of the cat cerebrum. I. The temporal region. Brain Res 51:1–22

Kiang NY-S, Morest DK, Godfrey DA, Guinan JJ, Kane EC (1973) Stimulus coding at caudal levels of the cat's auditory nervous system. I. Response characteristics of single units. In: Møller A (ed) Basic mechanisms in hearing. Academic Press, New York, pp 455–478

Kimura H, McGeer PL, Peng JH, McGeer EG (1981) The central cholinergic system studied by choline acetyltransferase immunohistochemistry in the cat. J Comp Neurol 200:151–201

Knight PL (1977) Representation of the cochlea within the anterior auditory field (AAF) of the cat. Brain Res 130:447–467

Kolb H, Normann RA (1982) A-type horizontal cells of the superior edge of the linear visual streak of the rabbit retina have oriented, elongated dendritic trees. Vision Res 22:905–916

Kölliker A (1896) Handbuch der Gewebelehre des Menschen. vol 2. Engelmann, Leipzig (2nd edition)

Kopsch F (1896) Erfahrungen über die Verwendung des Formaldehyde bei der Chromsilber-Impregnation. Anat Anz 11:727

Kudo M, Niimi K (1980) Ascending projections of the inferior colliculus in the cat: an autoradiographic study. J Comp Neurol 191:545–556

Kuhlenbeck H (1966) The central nervous system of vertebrates. vol I. Propaedeutics to comparative neurology. Karger, Basel

Kuhlenbeck H (1973) The central nervous system of vertebrates. vol III, Part 2. Overall morphologic pattern. Karger, Basel

Kuroda R, Murui H, Akagi K, Kamikawa K, Mogami H (1975) Efferent connections of the centromedian nucleus and the magnocellular part of the medial geniculate body in cats. Confin Neurol 37:120–127

LaVail JH, Winston KR, Tish A (1973) A method based on retrograde intraaxonal transport of protein for identification of cell bodies of origin of axons terminating within the CNS. Brain Res 58:470–477

Leibler L (1976) Monaural and binaural pathways in the ascending auditory system of the pigeon. Thesis, Massachusetts Institute of Technology

Leontovich TA (1975) Quantitative analysis and classification of subcortical forebrain neurons. In: Santini M (ed) Golgi centennial symposium, perspectives in neurobiology. Raven Press, New York, pp 101–122

Leppelsack H-J (1974) Funktionelle Eigenschaften der Hörbahn im Feld L des Neostriatum caudal des Staren. J Comp Physiol 88:271–320

Lippe WR, Weinberger NM (1973a) The distribution of click-evoked activity within the medial geniculate body of the anesthetized cat. Exp Neurol 39:507–523

Lippe WR, Weinberger NM (1973b) The distribution of sensory-evoked activity within the medial geniculate body of the anesthetized cat. Exp Neurol 40:431–444

Lorente de Nó R (1934) Studies on the structure of the cerebral cortex. II. Continuation of the study of the Ammonic system. J Psychol Neurol (Lpz) 46:113–177

Lund JS (1973) Organization of neurons in the visual cortex, area 17, of the monkey (*Macaca mulatta*). J Comp Neurol 147:455–496

Lund JS, Hendrickson AE, Ogren MP, Tobin EA (1981) Anatomical organization of primate visual cortex area VII. J Comp Neurol 202:19–46

Mannen HL (1960) Noyau fermé et noyau ouvert. Arch Ital Biol 98:333–350

McAllister JP, Wells J (1981) The structural organization of the ventral posterolateral nucleus in the rat. J Comp Neurol 197:271–301

Mehler WR, Feferman ME, Nauta WJH (1960) Ascending axon degeneration following anterolateral cordotomy. An experimental study in the monkey. Brain 83:718–750

Merzenich MM, Knight PL, Roth GL (1975) Representation of cochlea within primary auditory cortex in the cat. J Neurophysiol 38:231–249

Merzenich MM, Kaas JH, Sur M, Lin C-S (1978) Double representation of the body surface within cytoarchitectonic areas 3b and 1 in "S1" in the owl monkey (*Aotus trivirgatus*). J Comp Neurol 181:41–74

Merzenich MM, Kaas JH (1980) Principles of organization of sensory-perceptual systems in mammals. In: Stellar E, Sprague JM (eds) Progress in psychobiology and physiological psychology, vol 9. Academic Press, New York, pp 1–42

Mickle WA, Ades HW (1954) Rostral projection of the vestibular system. Am J Physiol 176:243–246

Middlebrooks JC, Dykes RW, Merzenich MM (1980) Binaural response-specific bands in primary auditory cortex (AI) of the cat: topographical organization orthogonal to isofrequency contours. Brain Res 181:31–48

Middlebrooks JC, Zook JM (1983) Intrinsic organization of the cat's medial geniculate body identified by projections to binaural response-specific bands in the primary auditory cortex. J Neurosci 3:203–224

Moore RY, Goldberg JM (1963) Ascending projections of the inferior colliculus in the cat. J Comp Neurol 121:109–135

Morel A, Imig TJ (1982) Frequency representation in the cat's medial geniculate body (MGB) and posterior complex (PO). Proc Soc Neurosci 8:349

Morest DK (1964) The neuronal architecture of the medial geniculate body of the cat. J Anat 98:611–630

Morest DK (1965a) The laminar structure of the medial geniculate body of the cat. J Anat 99:143–160

Morest DK (1965b) The lateral tegmental system of the midbrain and the medial geniculate body: study with Golgi and Nauta methods in the cat. J Anat 99:611–634

Morest DK (1965c) Identification of homologous neurons in the posterolateral thalamus of cat and Virginia opossum. Anat Rec 151:390–391

Morest DK (1969) The growth of dendrites in the mammalian brain. Z Anat Entwickl-Gesch 128:290–317

Morest DK (1971) Dendrodendritic synapses of cells that have axons: the fine structure of the Golgi type II cell in the medial geniculate body of the cat. Z Anat Entwickl-Gesch 133:216–246

Morest DK (1975a) Synaptic relationships of Golgi type II cells in the medial geniculate body of the cat. J Comp Neurol 162:157–194

Morest DK (1975b) Structural organization of the auditory pathways. In: Tower DB (ed) The nervous system. Human communication and its disorders vol 3. Raven Press, New York, pp 19–29

Morest DK, Morest RR (1966) Perfusion-fixation of the brain with chrome-osmium solutions for the rapid Golgi method. Am J Anat 118:811–832

Morest DK, Winer JA (to be published) The comparative anatomy of neurons: homologous neurons in the medial geniculare body of the opossum and the cat. Adv Anat Embryol Cell Biol

Morest DK, Kiang NY-S, Kane EC, Guinan JJ, Godfrey DA (1973) Stimulus coding at caudal levels in the cat's auditory system. II. Patterns of synaptic organization. In: Møller A (ed) Basic mechanisms in hearing. Academic Press, New York, pp 479–504

Morrison JH, Grzanna R, Molliver ME, Coyle JT (1978) The distribution and orientation of noradrenergic fibers in neocortex of the rat: an immunofluorescence study. J Comp Neurol 181:17–40

Nelson PG, Erulkar SD (1963) Synaptic mechanisms of excitation and inhibition in the central auditory pathway. J Neurophysiol 26:908–923

Niimi K, Matsuoka H (1979) Thalamocortical organization of the auditory system in the cat studied by retrograde axonal transport of horseradish peroxidase. Adv Anat Embryol Cell Biol 57:1–56

Niimi K, Naito F (1974) Cortical projections of the medial geniculate body in the cat. Exp Brain Res 19:326–342

Ogren MP, Hendrickson AE (1979) The structural organization of the inferior and lateral subdivisions of the *Macaca* monkey pulvinar. J Comp Neurol 179:147–178

O'Leary JL (1940) A structural analysis of the lateral geniculate nucleus of the cat. J Comp Neurol 73:405–430

Oliver DL (1982) A Golgi study of the medial geniculate body in the tree shrew (*Tupaia glis*). J Comp Neurol 209:1–16

Oliver DL, Hall WC (1978a) The medial geniculate body of the tree shrew, *Tupaia glis*. I. Cytoarchitecture and midbrain connections. J Comp Neurol 182:423–458

Oliver DL, Hall WC (1978b) The medial geniculate body of the tree shrew, *Tupaia glis*. II. Connections with the neocortex. J Comp Neurol 182:459–494

Orman SS, Humphrey GL (1981) Effects of changes in cortical arousal and of auditory cortex cooling on neuronal activity in the medial geniculate body. Exp Brain Res 42:475–482

Palmer LA, Rosenquist AC, Tusa RJ (1978) The retinotopic organization of lateral suprasylvian visual areas in the cat. J Comp Neurol 177:237–256

Papez JW (1929a) Central acoustic tract in cat and man. Anat Rec 42:60

Papez JW (1929b) Comparative neurology. Crowell, New York

Papez JW (1936) Evolution of the medial geniculate body. J Comp Neurol 64:41–61

Paula-Barbosa M, Feyo PB, Sousa-Pinto A (1975) The association connexions of the suprasylvian fringe (SF) and other areas of the cat auditory cortex. Exp Brain Res 23:535–554

Phillips DP, Irvine DRF (1981) Responses of single neurons in physiologically defined primary auditory cortex (AI) of the cat: frequency tuning and responses to intensity. J Neurophysiol 45:48–58

Poggio GF, Mountcastle VB (1960) A study of the functional contributions of the lemniscal and spinothalamic system to somatic sensibility. Central nervous mechanisms in pain. Bull Johns Hopkins Hosp 106:266–316

Poggio GF, Mountcastle VB (1963) The functional properties of ventrobasal thalamic neurons studied in unanesthetized monkeys. J Neurophysiol 26:775–806

Pontes C, Reis FF, Sousa-Pinto A (1975) The auditory cortical projections onto the medial geniculate body of the cat. An experimental study with silver and autoradiographic methods. Brain Res 91:43–63

Pritz MB (1974) Ascending connections of a midbrain auditory area in a crocodile, *Caiman crocodilus*. J Comp Neurol 153:179–198

Pritz MB (1980) Parallels in the organization of visual and auditory systems in crocodiles. In: Ebbesson SOE (ed) Comparative neurology of the telencephalon. Plenum Press, New York, pp 331–342

Ralston HJ III (1971) Evidence for presynaptic dendrites and a proposal for their mechanism of action. Nature 230:585–587

Ramon-Moliner E (1969) The leptodendritic neuron: its distribution and significance. Ann NY Acad Sci 167:65–70

Ramon-Moliner E (1970) The Golgi-Cox technique. In: Nauta WJH, Ebbesson SOE (eds) Contemporary research methods in neuroanatomy. Springer, Berlin Heidelberg New York, pp 32–55

Ramon-Moliner E, Nauta WJH (1966) The isodendritic core of the brain stem. J Comp Neurol 126:311–336

Ramón y Cajal S (1911) Histologie du système nerveux de l'homme et des vertébrés.Translated by L Azoulay. Instituto Ramón y Cajal el Consejo Superior de Investigaciones Çientificas, Madrid (1972 reprint)

Raymond J, Sans A, Marty R (1974) Projections thalamiques des noyaux vestibulaires: étude histologique chez le chat. Exp Brain Res 20:273–283

Reale RA, Imig TJ (1980) Tonotopic organization in auditory cortex of the cat. J Comp Neurol 192:265–291

Rioch D McK (1929) Studies on the diencephalon of carnivora. Part I. The nuclear configuration of the thalamus, epithalamus, and hypothalamus of the dog and cat. J Comp Neurol 49:1–119

Rose JE (1949) The cellular structure of the auditory region of the cat. J Comp Neurol 91:409–440

Rose JE, Woolsey CN (1949) The relations of thalamic connections, cellular structure, and evocable electrical activity in the auditory region of the cat. J Comp Neurol 91:441–466

Rose JE, Woolsey CN (1958) Cortical projections and functional organization of thalamic auditory system of cat. In: Harlow HF, Woolsey CN (eds) Biological and biochemical bases of behavior. University of Wisconsin Press, Madison, pp 127–150

Roucoux-Hanus M, Boisacq-Schepens N (1977) Ascending vestibular projections: further results at cortical and thalamic levels in the cat. Exp Brain Res 29:283–292

Rouiller E, de Ribaupierre Y, de Ribaupierre F (1979) Phase-locked responses at low frequency in the medial geniculate body. Hear Res 1:213–226

Russchen FT (1982) Amygdalopetal projections in the cat. II. Subcortical afferent connections. A study with retrograde tracing techniques. J Comp Neurol 207:157–176

Ryugo DK, Killackey HP (1974) Differential telencephalic projections of the medial and ventral divisions of the medial geniculate body of the rat. Brain Res 82:173–177

Ryugo DK, Weinberger NM (1976) Cortico-fugal modulation of the medial geniculate body. Exp Neurol 51:377–391

Ryugo DK, Weinberger NM (1978) Differential plasticity of morphologically distinct populations in the medial geniculate body of the cat during classical conditioning. Behav Biol 22:275–301

Saini KD, Leppelsack H-J (1977) Neuronal arrangement in the auditory field L of the neostriatum of the starling. Cell Tissue Res 176:309–316

Saini KD, Leppelsack H-J (1981) Cell types of the auditory caudomedial neostriatum of the starling (*Sturrus vulgaris*). J Comp Neurol 198:209–229

Scheibel ME, Scheibel AB (1958) Structural substrates for integrative patterns in the brain stem reticular core. In: Jasper HH, Proctor LD, Knighton RS, Noshay WC, Costello RT (eds) Reticular formation of the brain. Little, Brown, New York, pp 31–55

Scheibel ME, Scheibel AB (1966a) Patterns of organization in specific and non-specific thalamic fields. In: Purpura DP, Yahr MD (eds) The thalamus. Columbia University Press, New York, pp 13–46

Scheibel ME, Scheibel AB (1966b) The organization of the nucleus reticularis thalami: a Golgi study. Brain Res 1:43–62

Scheibel ME, Scheibel AB (1967) Structural organization of non-specific thalamic nuclei and their projection toward cortex. Brain Res 6:60–94

Scheibel ME, Scheibel AB, Davies TL (1972) Some substrates for centrifugal control over thalamic cell ensembles. In: Frigyesi TL, Rinvik E, Yahr MD (eds) Corticothalamic projections and sensorimotor activities. Raven Press, New York, pp 131–156

Schneider GE (1969) Two visual systems. Science 163:895–902

Somogyi G, Hajdu F, Tömböl T, Madarász M (1979) Types of thalamo-cortical relay neurons in the anteroventral nucleus of the cat. A combined horseradish peroxidase-Golgi study. Cell Tissue Res 196:175–179

Sousa-Pinto A (1973) Cortical projections of the medial geniculate body of the cat. Adv Anat Embryol Cell Biol 48 (2):1–42

Spreafico R, Hayes NL, Rustioni A (1981) Thalamic projections to the primary and secondary somatosensory cortices in cat: single and double retrograde tracer studies. J Comp Neurol 203:67–90

Tebēcis AK (1972) Cholinergic and non-cholinergic transmission in the medial geniculate nucleus of the cat. J Physiol 226:153–172

Tusa RJ, Palmer LA, Rosenquist AC (1978) The retinotopic organization of area 17 (striate cortex) in the cat. J Comp Neurol 177:213–236

Walker AE (1938) The primate thalamus. University of Chicago Press, Chicago

Wepsic JG (1966) Multimodal sensory activation of cells in the magnocellular medial geniculate nucleus. Exp Neurol 15:299–318

Whitfield IC, Purser D (1972) Microelectrode study of the medial geniculate body in unanesthetized free-moving cats. Brain Behav Evol 6:311–322

Winer JA (1977) A review of the status of the horseradish peroxidase method in neuroanatomy. Neurosci Biobehav Rev 1:45–54

Winer JA (1979) Neurons in the medial division of the medial geniculate body of the cat. Anat Rec 193:723

Winer JA (1981) Structure of laminae in the central nucleus of the inferior colliculus of the bat (*Antrozous pallidus*). Proc Soc Neurosci 7:58

Winer JA (1982) The stellate neurons in layer IV of primary auditory cortex (AI) of the cat: a study of columnar organization. Proc Soc Neurosci 8:1020

Winer JA (1984a) Anatomy of layer IV in cat primary auditory cortex (AI). J Comp Neurol 224:535–567

Winer JA (1984b) Identification and structure of neurons in the medial geniculate body projecting to primary auditory cortex (AI) in the cat. Neuroscience, in press

Winer JA (1984c) The human medial geniculate body. Hear Res, 15:225–247

Winer JA, Morest DK (1978) Morphology of neurons and axons in the dorsal nucleus of the medial geniculate body of the cat: study with the Golgi method. Proc Soc Neurosci 4:12

Winer JA, Morest DK (1979) What is a homology in the central nervous system? Golgi study of the opossum and cat medial geniculate body. Proc Soc Neurosci 5:147

Winer JA, Morest DK (1983a) The neuronal architecture of the dorsal division of the medial geniculate body of the cat. A study with the rapid Golgi method. J Comp Neurol 221:1–30

Winer JA, Morest DK (1983b) The neurons of the medial division of the medial geniculate body of the cat: implications for thalamic organization. J Neurosci 3:2629–2651

Winer JA, Morest DK (1984) Axons of the dorsal division of the medial geniculate body of the cat: a study with the rapid Golgi method. J Comp Neurol 224:344–370

Winer JA, Diamond IT, Raczkowski D (1977) Subdivisions of the auditory cortex of the cat: the retrograde transport of horseradish peroxidase to the medial geniculate body and posterior thalamic nuclei. J Comp Neurol 176:387–418

Winguth SD, Winer JA (1983) Architecture of layer II in cat primary auditory cortex (AI) and some ipsilateral cortico-cortical connections. Proc Soc Neurosci 9:954

Woolsey CN (1961) Organization of cortical auditory system. In: Rosenblith WA (ed) Sensory communication. MIT Press, Cambridge, pp 235 257

Woolsey CN (1972) Designated discussion. Brain Behav Evol 6:323–328

Woolsey CN (ed) (1981) Cortical sensory organization. Humana Press, Clifton

Subject Index

L. Heimer

The Human Brain and Spinal Cord

Functional Neuroanatomy and Dissection Guide

1983. 213 figures, mostly in color. XI, 402 pages.
ISBN 3-540-90741-6

„The text is simple, straight forward, concise, and free of jargon. ... this book is a useful medial neuroanatomy text and will undoubtedly be of considerable service to those who teach this subject. Careful consideration should be given to this book when texts for medial neuroanatomy are discussed by a teaching faculty."
Journal of Neuroscience Research

The Human Brain and Spinal Cord is a concise introduction to neuroanatomy and neuroscience written by one of the most respected neuroanatomy educators in the US. The text has been developed and refined from the author's teaching experiences in both the US and Europe. For the first time, a meticulously illustrated dissection guide is included, coordinated to the functional anatomy text to save students the expense and trouble of buying an additional neuroanatomy atlas. The book is divided into four parts. Part 1 explains terminology, brain and spinal cord development, and the meninges and cerebrospinal fluid. Part 2 is the dissection guide. Found here are outstanding, detailed illustrations which were created especially for this book. Text and illustrations are arranged to maximize the book's usefulness in the dissecting room. Part 3 ist an illustrated account of functional anatomy of the brain and spinal cord, with schematic drawings based on the illustrations found in Part 2. Part 4 covers the blood supply to the brain and spinal cord. An appendix on peripheral nerves is also included. Clinical correlations and case studies are provided throughout the book to emphasize the clinical relevance of the information.

Springer-Verlag
Berlin
Heidelberg
New York
Tokyo

Advances in Anatomy, Embryology and Cell Biology

Editors: F.Beck, W.Hild, J.van Limborgh,
R.Ortmann, J.E.Pauly, T.H.Schiebler

Volume 85
J.Altman

**The Development of the
Rat Spinal Cord**

1984. 126 figures. VIII, 166 pages
ISBN 3-540-13119-1

Volume 84
F.Hajós, E.Bascó

The Surface-Contact Glia

1984. 25 figures. VI, 81 pages
ISBN 3-540-13243-0

Volume 83
W.K.Schwerdtfeger

**Structure and Fiber Connections of
the Hippocampus**

A Comparative Study
1984. 40 figures. VI, 74 pages
ISBN 3-540-13092-6

Volume 82
H.Scheich, S.O.E.Ebbesson

**Multimodal Torus in the
Weakly Electric Fish Eigenmannia**

1983. 39 figures. VII, 69 pages
ISBN 3-540-12517-5

Volume 81
U.-F.Habenicht, F.Neumann

**Hormonal Regulation of
Testicular Descent**

1983. 39 figures. VI, 55 pages
ISBN 3-540-12439-X

Volume 80
J.Koebke

**A Biomechanical and Morphological
Analysis of Human Hand Joints**

1983. 50 figures. VI, 85 pages
ISBN 3-540-12438-1

Volume 79
S.F.Perry

Reptilian Lungs

Functional Anatomy and Evolution
1983. 32 figures. VII, 81 pages
ISBN 3-540-12194-3

Volume 78
G.Grün

**The Development of the Vertebrate
Retina: A Comparative Survey**

1982. 15 figures. VIII, 85 pages
ISBN 3-540-11770-9

Volume 77
E.Braak

**On the Structure of the
Human Striate Area**

1982. 44 figures. VI, 87 pages
ISBN 3-540-11512-9

Volume 76
P.Kugler

**On Angiotensin-Degrading
Aminopeptidases in the Rat Kidney**

1982. 88 figures. 96 pages
ISBN 3-540-11452-1

Volume 75
V.Grouls, B.Helpap

**The Development of the Red Pulp
in the Spleen**

1982. 37 figures. 80 pages
ISBN 3-540-11408-4

Springer-Verlag
Berlin
Heidelberg
New York
Tokyo